ELECTRICAL INSTALLATION

SNAGS and SOLUTIONS

ENLARGED EDITION

Recommendations on the application of the 14th edition of the IEE Regulations for the Electrical Equipment of Buildings

C. D. KINLOCH, D.F.H., C.Eng., M.I.E.E.
Assistant General Manager National Inspection Council for Electrical Installation Contracting

M. H. GOUGH
Assistant Editor — Contracting
Electrical Times

Published by
IPC Electrical-Electronic Press Ltd in collaboration with the National Inspection Council for Electrical Installation Contracting and Electrical Times

Technical material is based largely on information published in NICEIC Newsletters, relating to the 14th Edition of the IEE Regulations for the Electrical Equipment of Buildings.

1976 first edition by C. D. Kinloch and D. M. Williams

David Williams, Editor, Middle East Electricity, formerly with Electrical Times

Reprinted February 1977

1978 second, enlarged edition by C. D. Kinloch and M. H. Gough

ISBN 0 617 00185 5

Further copies of this book are obtainable from the publishers, and cheques, etc., with orders, should be sent to:

General Sales Department, Room 11, IPC Electrical-Electronic Press Ltd., Dorset House, Stamford Street, London, SE1 9LU.

Inclusive price is £2.30 and payment should be made to IPC Business Press Ltd.

NETHERTON AND WORTH LTD., LONDON AND TRURO

CONTENTS

INTRODUCTION *page 7*

1. INSTALLATION DESIGN *page 9*

2. WIRING SYSTEMS *page 21*

3. EARTHING AND BONDING *page 33*

4. ISOLATION AND PROTECTION *page 43*

5. LIGHTING *page 53*

6. SPECIAL ENVIRONMENTS *page 59*

7. INSPECTION AND TESTING *page 65*

8. INSTALLATION CHECK LISTS *page 73*

9. USEFUL PUBLICATIONS *page 81*

INDEX BY SUBJECT *page 85*

INDEX BY REGULATION NUMBER *page 88*

INTRODUCTION TO FIRST EDITION

For nearly 20 years, the Newsletters published by the National Inspection Council for Electrical Installation Contracting have included technical information on almost every aspect of good elecrical installation practice based on the requirements of the IEE Wiring Regulations.

While primarily aimed at approved electrical installation contractors, the information contained in the Newsletters has been invaluable to every one concerned with safe electrical installations, whether he be an apprentice, electrician, designer, consultant, architect, local authority engineer, etc.

The technical items contained in this book have been extracted from NICEIC Newsletters and have been revised and expanded to take account of the latest amendments to the 14th Edition of the IEE Wiring Regulaions. Most of the subjects covered are fundamental to the safe use of electricity and the advice given will still be generally relevant even though the wording and numbering of individual regulations will be different in future editions of the IEE Wiring Regulations.

The number of enquiries dealt with every day by NICEIC staff shows that there is a real need for clear and accurate advice on the correct application of the IEE Wiring Regulations. This book provides the answers to many common orpblems; it deserves to be widely read and to be kept for ready reference wherever electrical installations are designed and carried out.

ALAN PLUMPTON,

CHAIRMAN, (1976-77)

NATIONAL INSPECTION COUNCIL FOR ELECTRICAL INSTALLATION CONTRACTING

INTRODUCTION TO ENLARGED EDITION

The success of the first edition of Snags and Solutions clearly demonstrates that this publication has gone a long way towards filling a gap in the available literature on electrical installation practice.

This enlarged edition incorporates additional items selected from the Council's popular Newsletters, making it even more useful for everyone concerned with the correct application of the IEE Wiring Regulations.

The installation check lists, a popular feature in the original edition, remains as before but a list of addresses has been added showing where selected useful publications may be obtained.

GEORGE REID,

CHAIRMAN (1977-78),

NATIONAL INSPECTION COUNCIL FOR ELECTRICAL INSTALLATION CONTRACTING

1. INSTALLATION DESIGN

1.1 Meter tails for domestic installations

Selection of the appropriate rating for meter tails on installations which do not incorporate a consumer's main fuse can cause problems for contractors.

Regulation A.1(ii) calls for means of excess-current protection on every consumer's installation and Regulation A.3 states that this must consist of a fuse or excess-current circuit-breaker in each live conductor. But Regulation A.3 also contains an exemption whereby the means of excess-current protection may be omitted subject to two conditions being met.

One condition is that the main switch must be in the same enclosure as, or immediately adjacent to, the fuses or circuit-breakers controlling the sub-circuits. This may be achieved by using a consumer's control unit or a switch adjacent to the distribution board.

The other condition is that all cables on the incoming side of the main switch must have a current rating not less than the ratnng of the service fuses; in practice the cables referred to are the tails from the supply meter to the consumer unit, which are normally the property of the consumer.

The requirement causes problems because it is not always possible to ascertain the rating of the service fuse installed by the supply authority. It is recommended, therefore, that the size of the meter tails should be determined by reference to the load requirements of the installation, namely 16 sq mm up to 60A load or 25 sq mm up to 100A load.

1.2 Interconnections between switchgear

There are certain exemptions to the general requirement of Regulation A.10 that every conductor must be protected against excess current by a fuse or circuit-breaker of rating not exceeding the rating of the conductor. Seven items are listed as exemptions but, in addition, Regulations A.11 and A.12 also provide for departures from the general principle.

Regulation A.11 deals with cables interconnecting switchgear. If these do not exceed 2m in length, and have a current rating not less than the fuse or circuit-breaker which they supply, the fuse or circuit-breaker protecting them may have a rating of not more than twice the rating of the cable.

Regulation A.12 permits a further reduction of cable size in the case of composite switchboards and floor distribution boards associated with rising mains systems, provided that the cables are entirely enclosed in earthed metal. The cable sizes may be determined by the rating of the fuse or circuit-breaker supplied by the cables, subject to a minimum size of 4 sq mm.

None of these exemptions applies to meter tails where the consumer's main fuse or circuit-breaker has been omitted in accordance with the exemption to Regulation A.3.

1.3 Double-pole switches

Occasions when it is necessary to install a double-pole switch on a single-phase installation, other than as the main isolator for the whole installation, are listed below.

1. To control any appliance fitted with heating elements which can be touched, including silica-sheathed elements (Regulation A.58).

2. To control any installation in which neither pole is connected with earth, for example a circuit supplied through a transformer with a secondary winding having the centre tap earthed (Regulation A.9).

3. To control a step-up transformer (Regulation C.31).

4. To control a water heater with uninsulated elements immersed in water (Regulation C.50).

5. To control circuits installed in explosive atmospheres, such as circuits for petrol pumps (CP 1003).

6. For isolation of, and as a "fireman's switch" for, circuits supplying discharge lamps at voltages in excess of 600V to earth or 1,000V between conductors (Regulations G.11 and G.13).

1.4 Selection of cables

It is not always easy to ensure that all relevant factors have been taken into account when selecting cables for electrical installations. The following step-by-step approach should help.

Steps 1-5 apply to all types of installation covered by the IEE Wiring Regulations while steps 6-8 apply to multiple circuits using insulated cables in trunking.

1. Determine initially the following items:
(a) Type of cable (e.g. insulated, sheathed, armoured etc.)
(b) Method of installation (e.g. surface or enclosed)
(c) Full-load current for each circuit
(d) Single-phase or three-phase
(e) Class of excess-current protection (close or coarse)
(f) Ambient temperature
(g) Number of circuits in each group
(h) Length of run

2. Select the appropriate current rating table according to the type of cable.

3. Note the rating factors for the following:
(i) Class of excess-current protection
(ii) Ambient temperature
(iii) Number of circuits
Multiply these three factors together to determine the aggregate rating factor.

4. Select the current rating column which is appropriate to the method of installation and number of phases, and multiply the listed current ratings by the aggregate rating factor to obtain the actual ratings. Select one which is not less than the full-load current of the circuit and note what cable size is indicated in Column 1.

5. Note the listed volt drop in mV for the cable size selected and multiply by the full-load current (in amperes) and the length of run (in metres) and divide by 1,000 to obtain the volt drop (in volts) at full-load current.

Determine the overall volt drop by adding the volt drop for other parts of the circuit from the point of supply to each point of utilisation. If the total volt drop is less than 6V for single-phase circuits or 10V (line voltage) for balanced three-phase circuits, the size of cable selected is satisfactory. If the overall volt drop exceeds 6V or 10V respectively, select a larger cable size and repeat this volt drop calculation.

For p.v.c.-insulated cables in trunking, continue as follows:

6. Using Columns 1 and 3 of Table B.5M (or Table B.6M) determine the nominal overall diameter of the cable selected. Calculate the overall cross-sectional area of the cable from this using the following formula:

$$\text{Cross-sectional area} = \text{Diameter}^2 \times \frac{\pi}{4}$$

$(\pi = 3.142)$

7. Determine the aggregate cross-sectional area of the cables by adding the cross-sectional areas of all the individual cables.

8. Multiply the aggregate cross-sectional area of the cables by 2.22 to determine the minimum cross-sectional area of trunking needed to contain these cables in compliance with the 45 per cent space factor requirement of Regulation B.118.

1.5 Cable ratings and conditions of use

Current ratings quoted by some cable manufacturers for p.v.c./s.w.a./p.v.c. cables appear to permit larger currents for a given cable size than the IEE Wiring Regulations and this is a source of confusion for some people.

Cable manufacturers' ratings are often based on the Electrical Research Association (ERA) Report 69-30 Part III, which includes ratings for these cables "in air". This applies to cables installed singly, fixed to a vertical wall but with a clear space of 20mm between the cable and the wall, and having close excess-current protection.

These "in air" conditions correspond exactly with the "defined conditions" of the IEE Regulations, but the current ratings tabulated in the ERA report are for an ambient temperature of 25degC and those in the IEE Regulations are for 30degC. When the appropriate correction factor for ambient temperature is applied, both sets of ratings will be found to be identical.

It should be noted that for groups of cables, the "unenclosed" ratings should be used, together with the appropriate group factor. "In air" or "defined conditions" ratings can only be used if the horizontal spacing between circuits is more than six times the overall diameter of individual cables or 150mm, whichever is the lower, and the vertical clearance is not less than 150mm.

"Defined conditions" can never be applied to multiple cable runs in small trenches, even when left exposed, and when a trench is filled with dry sand the effective rating is further reduced to about three quarters of the "unenclosed" rating.

This underlines the importance of quoting all relevant information when seeking information on cable ratings from cable manufacturers.

1.6 Fixed appliances on ring circuits

Although it is permissible to connect fixed appliances such as water heaters, storage heaters or radiant heaters to a general-purpose domestic ring circuit, provided the total connected load does not exceed 7kW and individual loads do not exceed 3kW, it is bad practice to do so in the case of loads having little or no diversity.

The connection of a substantial fixed load to a ring circuit may leave little or no margin for the connection of other loads and overload conditions are likely to occur if the consumer makes full use of the ring circuit facility. It is recommended that separate final sub-circuits should be provided for each heavily loaded appliance in regular use.

1.7 Wiring for storage heaters

Two errors which must be avoided when installing modern "high capacity" storage radiators rated 3.375kW are the use of 1.5sq mm supply conductors and connection through 13A ring circuit accessories such as fused connection units.

The 3.375kW storage radiator has a current rating of 14A at 240V. Reference to Column 3 on Tables 1M and 3M in the IEE Wiring Regulations will show that the current rating for two single-core cables or one twin cable enclosed in conduit (or embedded in plaster) with coarse excess-current protection is 13A for 1.5 sq mm conductor and 18A for 2.5 sq mm conductor. Clearly conductors must be no less than 2.5 sq mm unless close excess-current protection is provided by cartridge fuses or miniature circuit-breakers.

While it may be argued that, from column 7 of Table 3M, a rating of 15 A is given for a 1.5 sq mm twin cable clipped direct to a surface, it must be remembered that de-rating is necessary if cables are bunched at any point, for example near the distribution board or consumer unit.

On the second point mentioned above, appliances rated more than 3kW are still occasionally found connected through 13A fused connection units, which leads to over-heating of the accessory and operation of the fuse. Storage heaters of this rating should be connected to individual final sub-circuits, with 15A or 20A switches adjacent to each heater.

Regulation A.62 now permits the switch for a comprehensive heating installation to be installed in a separate room, even in domestic premises, and it is therefore now possible to dispense with the switches adjacent to the heaters and to rely on a single switch (e.g. the switch in a consumer unit exclusive to the heater circuits) for isolation of all the heaters provided that individual on/off control is incorporated in each appliance or is not required by the user.

1.8 Final sub-circuits for Electricaire

Electricaire heaters should be controlled at the origin of the circuit by a separate single-pole and neutral switch-fuse rather than one fuse-way of a composite consumer control unit.

Problems have arisen where contractors have supplied a 10kW Electricaire unit (current rating 42A at 240V; 43.5A at 230V) through the 45A fuse-way on a composite control unit and then taken advantage of the close excess-current protection afforded by a BS 1361 fuse cartridge to reduce the size of the final sub-circuit conductors. Overheating has occurred at the termination, aggrevated by the close proximity of the other final sub-circuit fuse-ways. This does not normally occur when the 45A fuse-way is used, as intended, for a cooker circuit due to the diversity factor applying to cooking appliances.

An isolating switch may also be necessary adjacent to the Electricaire unit. Where the unit is "built-in" or enclosed within a cupboard, where relatively high ambient temperatures will prevail, switchgear intended for normal ambient temperatures and having a current rating close to the full-load current of the Electricaire unit should not be positioned in the enclosure. If it is essential to mount switchgear in conditions of high ambient temperature, advice should be obtained from the switchgear manufacturer regarding possible de-rating of the equipment.

Account should also be taken of the rating factors for ambient temperature when selecting cables for installation within the Electricaire enclosure.

1.9 Bell transformers

Certain small bell transformers do not have the provision for enclosure of cable cores that is necessary for compliance with Regulation B.6. On some types the mains-voltage terminals are exposed, leading to a breach also of Regulation 1(c).

It is particularly important to ensure safe installation of bell transformers because an isolating switch is not required by Regulation A.58.

It is recommended, therefore, that only bell transformers to BS 3535 Section B should be used. These transformers are double-insulated and there is no need to earth the secondary circuit or the core; most other types of bell transformer must be earthed in accordance with Regulation D.8.

1.10 Wiring for gas central heating

Isolation of the electrical supply to a gas central heating boiler should be provided by means of a three-pin plug and socket-outlet if the installation is to satisfy the requirements of the Confederation for the Registration of Gas Installers (CORGI) although the IEE Regulations would also permit the use of a switched spur connection unit or double-pole switch.

The Quality Standard Inspectors of CORGI, an organisation which carries out periodic inspection of its registered installers' work in a similar way to the NICEIC, assess installations by reference to British Standard Code of Practice CP 332: "Selection and installation of town gas space heating". This recommends that external wiring be in accordance with the IEE Wiring Regulations and that a three-pin plug and socket-outlet should be provided for isolation.

In addition every care must be taken to ensure that the earthing of the equipment is satisfactory and that overload protection is provided. A warning label should also be fixed in an appropriate position stressing that the appliance must be isolated from the electricity supply before normal servicing is carried out.

1.11 Timer-controlled extract fans

Special attention should be paid to the means of isolation provided with extract fans used for ventilation of windowless bathrooms and toilets in blocks of flats, hotels etc.

These fans usually start when the light is switched on and continue to run for a timed period after it has been switched off. Therefore, in addition to the supply controlled by the lighting switch, they require a maintained supply to operate the fan after the light has been switched off.

Some fans of this type incorporate means of isolation for both supplies, for example by a switch or plug-in link which completely isolates the unit on removal of the cover. If such isolation is not fitted by the manufacturer, separate means of isolation must be provided by the contractor for compliance with Regulation A.58 preferably by means of a switch arranged to isolate both supplies simultaneously but separate from the switch controlling the light.

1.12 Temperature characteristics of p.v.c. conduit

The temperature characteristics of rigid non-metallic conduit should be examined carefully in conjunction with the requirements of Regulations B.102-104 before designing or carrying out an installation in this type of conduit.

The note to Regulation B.102 gives limiting values of temperature, namely 60degC maximum working temperature and –5degC minimum ambient temperature, but with a lower limit of –25degC for Type B conduit to BS 4607 Part 1. Furthermore, the full current ratings of 85degC rubber-insulated cables cannot be used; only the ratings for p.v.c.-insulated cables are allowed.

Where runs of non-metallic conduit are required in locations subject to variations in temperature, expansion couplings and saddles must be used to allow for longitudinal expansion and contraction, as required by Regulation B.104. For a given change of temperature, p.v.c. conduit expands about five times as much as steel conduit; for example, a 3m length of p.v.c. conduit installed at an ambient temperature of 5degC will increase in length by 2.5mm when the temperature of the conduit rises to 30degC.

Where such provision has not been made, for example when conduit has been installed along purlins below a lightly-constructed roof in a factory or farm building, saddles have been known to be torn from their fixings or conduit to bow or fracture.

1.13 Isolation of motors

Requirements for isolation of electric motors are not always clearly understood. The following points should be noted:

1. A stop button, intended as a means of stopping a machine in accordance with Regulation A.63, is not acceptable as the means of isolation required by Regulation A.64(iii). The isolator must cut off voltage from the motor and all associated apparatus, including the starter and any other control equipment.
2. If possible, the isolator should be near the motor but where it is located remote from the motor, a padlock or other locking device is required to secure the isolator in the "off" position while work is being carried out on the motor or associated equipment. Alternatively, Regulation A.64(iii) permits the installation of a second isolator, adjacent to the motor, instead of a lock on the main isolator.
3. Where a second isolator is installed in the connection to slip-ring, star-delta start or large direct-on-line motors, it should be interconnected with a no-volt trip on the starter to prevent damage to the motor caused by closure of the local isolator with the starter in the "run" position. For smaller direct-on-line machines the local isolator can be inserted in the connection to the motor.

1.14 Stopping of motors

A remote stop button, or multiple stop buttons, may be used in addition to the stop button on a motor starter, to meet Regulation A.63 which requires provision of a means of stopping a motor "so placed as to be easily operated by the person in charge of the motor." Where such remote buttons are used, Regulation A.64(ii) calls for self-latching devices or other means of ensuring that all stop buttons must be re-set before the motor can be restarted.

1.15 Emergency lighting

Requirements for the provision of emergency lighting in most types of premises other than domestic, cinemas and certain specified premises for entertainment, are set out in the British Standard Code of Practice BS 5266 Part 1.

One of its requirements is that the wiring of emergency lighting installations must be segregated from the wiring of any other circuits. This applies to complete new installations where the system is supplied from a central battery or generator but not to individual self-contained luminaires incorporating battery, charger and control unit, when these are connected to the normal lighting supply source.

Self-contained luminaires require local means of isolation, however, for compliance with Regulation A.58. As it would be undesirable to provide a normal switch, it is recommended that use be made of either a key switch, a clock connector point or a ceiling rose of the type incorporating a socket-outlet.

There is no special problem regarding isolation with the separate installation of a centralised system because Regulation A.62 permits the switch, or switches, to be installed "in a separate room".

Before commencing work on an emergency lighting installation, a contractor should consult with the local enforcing authority to ensure that the system proposed is acceptable.

1.16 Fire alarm systems

Correct installation of electrical fire alarms is vital. Cases have occurred where false alarms have been so frequent that they have been ignored, and where incorrect connection of battery chargers has led to their being switched off, unplugged from a socket-outlet, or isolated as a result of somebody "borrowing" the fuse-link.

Recommendations for connection of the power supply are set out in the British Standard Code of Practice CP 1019: "The installation and servicing of electrical fire alarms", which should be in the possession of every electrical contractor.

The Code says that the supply equipment should be exclusive to the fire alarm system and should comprise a secondary battery and mains-operated charger, although in small buildings a direct mains-operated system or a primary battery may be acceptable.

Any connection to the mains supply should be through a switch-fuse painted red and labelled "Fire Alarm — Do Not Switch Off". It may be desirable to enclose the switch-fuse in a breakable glass fronted box and also provide a neon indicator lamp to show "mains on".

The point of connection to the mains should be selected to ensure continuity of supply. This normally precludes the use of one way in a distribution board which is used to supply other circuits. The switch-fuse should be connected directly to the consumer's main busbar or the equivalent point in small installations.

If the consumer's main switch is likely to be switched off at night, the switch-fuse for an automatic fire alarm system should be connected to the supply side of the main switch, as envisaged by Note 3 to Regulation A.1.

The switch-fuse should be suitable for the prospective fault current at the point of connection. For supplies of 200A and above, only switch-fuses incorporating h.b.c. fuses to BS 88 should be used to ensure compliance with Regulation A.6.

Contractors should also take steps to ensure that fire alarms will be maintained in working order and tested regularly in accordance with the Code of Practice. If possible, they should offer maintenance contracts on the systems they install.

1.17 Circuits below 240V

Circuits operating at 25V or 50V (extra-low voltage) or 110V are still governed by the IEE Wiring Regulations, which apply down to 15V between conductors or to earth. Despite the reduction in voltage, and corresponding increase in current for a given power rating, Regulations A.25 and A.26 still apply to the number of points that can be connected to a final sub-circuit.

For circuits rated at 15A or less, Table A.2 applies and so a 15A circuit may supply one 15A socket-outlet or three 5A socket-outlets. A circuit exceeding 15A rating may supply only one socket-outlet unless fused plugs are used in conjunction with BS 196 socket-outlets and these are connected to ring or radial circuits installed in accordance with Regulations A.43-55.

Sub-circuit cables must have the appropriate current rating indicated by the current rating tables. Due account must be taken of the rating factors for class of excess-current protection, ambient temperature and grouping. For voltages above 50V it is also essential to check the voltage drop. The permitted limit is 2.5 per cent of the circuit voltage, e.g. 2.75V at 110V.

Where the additional voltage is derived from a transformer, one point on the secondary winding must be earthed except as permitted by the exemptions to Regulation D.8 . If conditions permit omission of the earth connection, or if the centre-tap of the transformer secondary is earthed, Regulation A.9 requires double-pole switches and double-pole excess-current protection on the secondary side of the transformer.

1.18 Socket-outlets for voltages below 240V

When using socket-outlets for voltages below 240V, for example 25V and 50V in Band 1 (extra-low voltage) and 110V in Band II (low voltage). It is preferable to use BS 196 or BS 4343 types rather than BS 546 (2A, 5A and 15A round-pin) or BS 1363 (13A flat-pin) to avoid confusion with socket-outlets used at normal mains-voltage.

BS 4343 plugs and socket-outlets are dimensionally identical to European CEE 17 types and should be used where possible but BS 196 types may be necessary on extensions to existing installations. Both British Standards make provision for different keying arrangements to prevent insertion of a plug into a socket-outlet connected to a different voltage; it is essential to install the correct pattern for each voltage supply as indicated in the manufacturers' catalogues.

1.19 Voltages exceeding 250V

Where a voltage exceeding 250V exists between terminals or other live parts in an enclosure, Regulation A.17 calls for a clearly visible warning of the maximum voltage present. Regulation A.19 has a similar requirement where the live parts are in separate enclosures in reach of each other.

Because the term "medium voltage" is no longer recognised, the warnings must indicate the actual voltage, e.g. "415 Volts". There is no need to include the word

"Danger" but, to meet the requirements of Regulations A.19, it may be convenient to state "415 Volts between adjacent enclosures."

With industrial locations, it should be appropriate to fix suitable labels to the outside of busbar chambers, switches and distribution boards for example. With domestic installations, it is preferable to take advantage of the provision in Regulations A.16 and A.19 for the warning to be visible "before access is gained to live parts" and thus place it inside the enclosure such that it is seen as soon as the enclosure is opened.

1.20 Cables sizes for 30A circuits

To comply with the general requirements of Regulation A.10, the current rating of any cable must not be less than the rating of the fuse or circuit-breaker at the origin of the circuit (taking into account the class of excess-current protection provided).

Where 30A rewireable fuses (coarse excess-current protection) are installed, the minimum cable size is 6 sq mm as shown in Tables 1M and 3M, but 4 sq mm cables would be acceptable where 30A cartridge fuses or miniature circuit-breakers (close excess-current protection) are fitted.

The sizes quoted are correct for any 30A circuit, whether on the surface or enclosed, but must be increased when circuits are bunched, by applying the rating factors given below the current rating tables.

1.21 Assembly of metaclad switchboards

When switchgear is assembled on site using a busbar chamber, switch fuses or fused switches and trunking, it is often necessary to provide spacers between individual units to allow clearance for bolt heads or to obtain correct alignment.

As such spacers form part of the enclosure for non-sheathed cables, they should be non-combustible. When purpose-made components are not supplied by the switchgear manufacturer, it is recommended that spacers be cut from resin-bonded paper-laminted sheets complying with BS 5102; suitable materials are available under trade names such as Paxolin and Tufnol.

Items of switchgear are often supplied with temporary covers made of soft wood, which are only intended to provide protection against mechanical damage and ingress of foreign bodies before final assembly. Such wooden covers are not suitable for use as spacers and should be discarded.

1.22 Fire Prevention

The IEE Wiring Regulations are intended to ensure safety from fire and shock and the following paragraphs draw attention to some of the requirements relating specifically to preventing fires.

In any installation, heating can occur at the termination of conductors or at joints between conductors. Regulations B.60–78 are intended to ensure satisfactory electrical and mechanical conditions including enclosure in non-combustible materials (Regulation B.69).

Regulations C.11–16 deal with apparatus liable to cause overheating. In general, heat-producing equipment should be made and installed in accordance with a British Standard to comply with Regulation C.12(i). Where this is not possible, account must be taken of Regulation C.12(ii). Adequate ventilation is essential and soft wood or other combustible material must be protected by non-combustible material or spaced at distances not less than those shown in the diagram attached to this regulation.

Filament lamps must be placed or guarded so as to prevent ignition of flammable materials (Regulations C.15); shades should be capable of withstanding the heat of the lamp and it is important, therefore, that the wattage of the lamp does not exceed the recommendations of the manufacturer of the shade or luminaire.

Care is also necessary where lamps are installed in stair cupboards and roof spaces while Regulations J.17 and J.18 state the requirements for luminaires in caravans.

1.23 Installation of conduit for distribution between buildings

The use of metal conduit or pipe as a system for distribution between buildings is permissible provided the following methods and types of cable are used.

1. Heavy-gauge steel conduit, galvanised steel pipe or non-metallic conduit may be buried underground and used for protecting and enclosing either p.v.c.-insulated and sheathed cables or rubber-insulated cables having an oil-resisting and flame-retardant or h.o.f.r. sheath.

Steel conduit or pipe must not be used as an earth-continuity conductor but must be earthed in accordance with Regulation D.3 and buried at a depth of at least 500mm.

On farm or horticultural installations, however, the depth should not be less than 600mm to avoid risk of damage by agricultural implements; in some locations, depths of up to 1m may be required (Regulation K.20 and Note).

2. Heavy-gauge, hot-dipped zinc-coated steel conduit to BS31 or BS4568 or steel pipe systems complying with Regulations B97-100, may be used on exterior surface walls, boundary walls and the like for distribution between buildings but they must not be installed on fences.

3. Heavy-gauge hot-dipped zinc-coated steel conduit to BS31 or BS4568 enclosing either p.v.c.-insulated and sheathed cables or rubber-insulated cables having an oil-resisting and flame-retardant or h.o.f.r. sheath may be used, without intermediate support, for spanning between buildings provided that the conduit is fixed securely at the ends of the span, is not jointed, and is earthed in accordance with Section 'D'.

The maximum length of span shall not exceed 3m and the minimum height above ground level shall not be less than 3m. It should be noted, however, that this method of insulation is not permitted on farm and horticultural installations (Regulation K.22).

2. WIRING SYSTEMS

2.1 Protecting cables under floors and above ceilings

Since the April 1976 amendments to the IEE Wiring Regulations, the requirements of Regulation B.26 for protection of cables under floors have been extended to apply to cables in ceiling voids. Uncertainty has been expressed about the requirement when floor and ceiling surfaces are supported by battens above or below solid floor slabs.

The essential requirement is that "Where cables are installed under floors or above ceilings they shall run in such positions that they are not liable to be damaged by contact with the floor or the ceiling or their fixings."

In the case of wooden joists, where it is possible to do so without weakening the joists, this requirement should be met by drilling the joists to give a 50mm clearance from the top or bottom as appropriate. Alternatively, where it is necessary to lay cables in notches, the regulation is met by protecting the cable by steel conduit or equivalent mechanical protection (not rigid p.v.c. conduit).

Where thin battens are installed, the 50mm clearance cannot be achieved and so mechanical protection must be provided wherever cables are liable to damage by the floor or ceiling fixings, i.e. where the cables cross the battens.

It is hoped that architects and other designers will note this requirement and specify battens of sufficient depth to permit the use of 20mm conduit at least where cables cross the battens. Alternatively, consideration should be given to use of a complete steel conduit system for those parts of an installation fixed to either side of a solid floor slab.

2.2 Cable damage by vermin

Repeated cases of damage to p.v.c. cables by mice and rats are reported by electrical contractors. This damage occurs not only at the point where a cable crosses a "rat run" but where cables pass through holes in joists and over considerable lengths of cable where rodents have gnawed away the sheath and insulation, resulting in exposure of the conductors.

Wherever damage by rodents is liable to occur, for example in farm buildings or in the roof voids and cellars of older properties, the wiring system must be capable of resisting such damage and should be selected from the following:

1. Steel conduit or trunking
2. Rigid p.v.c. conduit or trunking
3. Mineral-insulated metal-sheathed cables
4. Steel wire armoured cables
5. PVC-sheathed cables having a galvanised steel covering
6. Steel capping to protect short exposed runs of p.v.c. insulated and sheathed cables.

Manufacturers of p.v.c. cables say there is no wholly effective means of prevention. If repellants were incorporated in the cable sheath they would almost certainly be toxic and so unsuitable for handling in cable factories and might also be a danger to people handling the finished cable.

2.3 Expanded polystyrene loft insulation

Enclosure of p.v.c. cables in conduit or channelling is recommended where it is known that expanded polystyrene granules are to be used for loft insulation as an alternative to glass fibre or vermiculite.

Where these expanded polystyrene granules are in contact with p.v.c.-sheathed cables, some of the plasticiser in the p.v.c. compound may migrate from the cable sheath to the polystyrene and form a tacky substance with some colour staining.

Investigations have shown that this migration may result in hardening and shrinkage of the p.v.c., which could lead to cracking if the cable is disturbed. There is, however, no change in the electrical properties of the p.v.c. and no significant change in the fire risk.

2.4 Temperature rise in m.i.m.s. cable

When mineral insulated metal sheathed cables are installed adjacent to ordinary p.v.c. cables, care must be taken that temperature rises in the m.i.m.s. cables are not so high as to cause deterioration in the p.v.c insulation. This applies to m.i.m.s. cables bunched in air with p.v.c. insulated and sheathed cables as well as to m.i.m.s. and p.v.c.-insulated cables installed together in trunking.

If the m.i.m.s. cables are loaded at currents no greater than those shown in Tables 13M and 14M, and the appropriate rating factors for ambient temperature and grouping are taken into account, no problems are likely to arise.

Use of the higher ratings given in Tables 15M and 16M for bare m.i.m.s. cables not exposed to touch would result in unacceptable temperature rises when the cables are installed adjacent to p.v.c. cables.

2.5 Voltage surges and m.i.m.s. cable

Mineral insulated metal sheathed cable may require surge voltage protection when used to supply certain types of inductive load. Surge voltages, generated when these loads are switched off, can be far in excess of the normal supply voltage and result in breakdown in the cable unless surge suppressors are fitted.

The principal sources of troublesome surges are (i) three-phase star-connected small power motors, particularly those used in heating and ventilating systems, (ii) fluorescent lighting circuits with a series-connected power factor correction capacitor or where there is no capacitor at all — particular offenders are 2,400mm 125W; 1,500mm 50W; 600mm 20W and miniature fluorescent lamps, (iii) motor control gear.

Where lighting control gear is mounted separately from the lamp, as in high pressure sodium and metal halide discharge lamps, and some mercury and fluorescent systems, surges are inevitable in the cables between the control gear and the lamps. Consequently, m.i. cables should never be used for this purpose, but they may be used for the supply to the control gear.

Contractors are advised to have the necessary suppressors fitted to equipment before delivery but where this is not possible, suitable surge suppressors should be obtained from the m.i.m.s. cable manufacturer.

Surge voltages do not normally cause trouble with other types of cable but they may occasionally be responsible for certain failures of switchgear or other equipment.

2.6 Terminating multi-core cables

Unused cores of multi-core cables should not be left "floating" in a box or could make the other end of the conductor live, where this is not expected.

Although it is not strictly a circuit conductor, the unused core should be correctly terminated or insulated at both ends to eliminate the possibility of contact with a live terminal. This would "prevent danger so far as is reasonably practicable" in the spirit of Regulation 1.

2.7 Terminals of socket-outlets

The variations in positioning of the live, neutral and earth terminals between different makes of socket-outlet, and even between different models by the same manufacturer, can create a hazard when an electrician who is accustomed to one type is called upon to install a different type.

British Standards specify the relative location of the contacts but the disposition of the terminals is not laid down so that it can be arranged to suit the ease of installation or manufacture, or just the whim of the designer. The function of the terminal must, of course, be marked but this does not entirely overcome the problem if terminals are transposed on two different types of socket-outlet.

If an earth-loop impedance test is carried out on every socket-outlet after completion of an installation any reversal of polarity will be revealed.

2.8 Use of cord grips

Regulation B.61 requires that cable terminations be made without imposing mechanical strain on the terminal. To enable installers to meet this requirement, cord-grip devices are incorporated in accessories such as ceiling roses, flexible cord outlet plates and cooker control units. It is essential that these devices are correctly used.

2.9 Compression-type connectors

Compression-type, or "crimp-type", connectors are acceptable for terminations and joints in compliance with Regulations B.61 and B.73, but it is essential that the correct tool is used for securing them. Reliance on tools such as ordinary pliers will result in joints having unduly high resistance and, consequently, in failure of the connection.

BS 4579 covers compression-type connectors ranging in size from those suitable for 1 sq mm cables to those suitable for 1,000 sq mm cables. Hand tools are available for the smaller sizes but an hydraulic tool is required for larger sizes to ensure an effective crimp.

2.10 Moulded connector blocks

Moulded connector strips and block are often used to make connections between cables in joint boxes and elsewhere. These items are only acceptable if the insulating portion is of a suitable material.

Far too much use is made of connectors shrouded in polyethylene or other materials with a relatively low melting point, particularly for connecting ceiling-mounted luminaires (see also **5.6**).

Connectors should be chosen with care so that they will withstand the maximum temperature likely to occur in the enclosure in which they are used.

Regulation B.74 states that "Every joint in a cable shall be provided with insulation not less effective than that of the cable cores and shall be protected against moisture and against mechanical damage." Clearly a connector shrouded in p.v.c. would satisfy this regulation when used for jointing p.v.c.-insulated cables, and connectors shrouded in thermosetting plastics or porcelain would be even more effective, but thermoplastics materials having a lower melting point than p.v.c. or poorer combustibility characteristics would not be satisfactory, even though the insulation resistance characteristics might be superior.

Where cables having insulation of dissimilar characteristics are jointed, for example general-purpose p.v.c. and 150degC rubber, the insulation of the joint must be related to the higher grade of the two insulating materials.

2.11 Borrowed neutrals

The practice of "borrowing" a neutral conductor from another circuit because it happens to be conveniently near to an outlet or switch position, is a deviation from Regulation A.24 and is dangerous since it can lead to serious risk of shock for anyone carrying out tests on the installation or making subsequent alterations or additions.

Instances are sometimes found where there are fewer neutral conductors than the number of circuits fed from a distribution board, indicating that a neutral may have been "borrowed" from another circuit, but the more common situation is in dwellings of two or more floors with two-way switching for the landing light.

If the feed for the landing light is derived from the ground floor lighting circuit, it is essential that the neutral connection is also made to the same circuit and *not* to the circuit for the lighting on the upper floor.

2.12 Neutrals at distribution boards

Identification of neutral conductors at a three-phase distribution board can be difficult if the arrangement of terminals on the neutral bar does not correspond exactly with that of the line terminals as envisaged by Regulation A.24. This can lead to danger if, as a result, an incorrect neutral is disconnected.

Some manufacturers provide a neutral bar chart in addition to the chart for fuseways. Although Regulation A.15 only calls for completion of the fuseway chart, a clear indication of the neutral connections is helpful and should always be provided when the layout of the neutral bar differs from the layout of the fuseways.

2.13 Slotted trunking and ducting

Semi-enclosed or slotted trunking and ducting does not satisfy the IEE Wiring Regulations' requirements for enclosure of non-sheathed cables.

This type of trunking and ducting is primarily intended for panel wiring and telecommunications applications and should not be used for installation work, except for containing sheathed cables.

2.14 Backless skirting trunking

"Backless" skirting trunking, which uses the building structure as part of the cable enclosure, does not satisfy Regulations B.28, B.69 and B.111 when used to carry non-sheathed cables and mounted on combustible surfaces such as soft wood.

Sheathed cables must be used where this type of trunking is installed on combustible surfaces.

Where the trunking is used for telecommunications circuits, as well as power circuits, care must also be taken to ensure adequate segregation as required by Regulations B.45–52.

2.15 Joints in trunking

When making up bends or connectors for trunking systems on site care should be taken to clean mating surfaces and to ensure the same degree of mechanical strength as with manufactured items. Cases have been noted where the only continuity path was provided by a single bolt or self-tapping screw.

Regulation D.29(v) states that "All joints in earth-continuity conductors shall be mechanically sound, electrically continuous and protected, where necessary, against corrosion."

2.16 Rivets for jointing of trunking

Techniques borrowed from another trade should be examined carefully for suitability before adoption for electrical installations.

While "pop rivets" may ensure a sound mechanical joint when used for jointing lengths of cable trunking, their ability to carry large earth-fault currents, when the trunking is used as an earth-continuity conductor, is doubtful. The area of contact made by these rivets is small, even when several are used in a joint, and some have a relatively low melting point due to the nature of the alloy used.

Many trunking manufacturers provide copper links for compliance with the earth-continuity requirements of BS 4678 for steel trunking; it is most important that these links are installed correctly, especially when rivets are used.

2.17 Corrosion in conduit systems

Dampness can affect conduit systems internally and externally and there are certain aspects of corrosion prevention which are sometimes overlooked.

It is not always appreciated that cutting the thread breaks the protective finish on corrosion-resistant conduit such as galvanised conduit. Immediate use of a suitable paint is essential to prevent rusting.

Internal corrosion can occur even in nominally dry situations, causing rapid deterioration and seriously affecting the function of the conduit as an earth-continuity conductor. To deal with this problem Regulation B.96 calls for the provision of ventilation and drainage points for condensed moisture, in conduit systems not intended to be gas-tight.

Suitable drainage outlets are provided by drilling 3mm diameter holes in the underside of draw-in boxes, switch boxes, enclosed luminaires etc. Not only will this permit the condensation to run out of the conduit system but it will also allow free circulation of air within the conduit, to permit drying out.

Conduit runs most susceptible to internal condensation are those subject to rapid temperature fluctuations, but it is good practice to provide suitable ventilation for all conduit systems whatever their location except, of course, for gas-tight or flameproof installations.

2.18 Conduit drops to motors in factories

Final connections to motors in factories frequently present problems because the supply has to come from overhead and there are no walls or other structural features to provide support for conduit drops.

It is not unusual to see conduit drops of 5m or more with one right-angle set and often a second, more open, set for which the only support is provided by the conduit entries at each end. At the least, this is a deviation from Regulation B.89 which requires secure fixing and protection from mechanical damage. If the conduit is relied upon for earth continuity this could also represent a deviation from D.29(v) which requires all joints in earth continuity conductors to be mechanically sound and electrically continuous — a knock from a ladder, vibration, or use of the conduit by the operator as a hand-hold, could all cause slackening of conduit joints leading to a high resistance or open-circuit earth-continuity path.

Where there is doubt about the reliability of earth-continuity paths, it is recommended that a separate earth-continuity conductor should be run in the conduit to avoid the deviation from D.29(v).

An alternative approach to the situation is to use a four-core flexible cable or cord, preferably metal-sheathed, to supply the machine, with only a short length of conduit at the lower end. This conduit should be firmly fixed to the machine, terminating with a ring bush or cable gland, and rising to about 2m above floor level. This method will accommodate discrepancies of one or two metres in the relative locations of the machine and supply point, and will provide mechanical protection adjacent to the machine.

2.19 Incorrect conduit fittings

Unsatisfactory earth-continuity in conduit systems is sometimes caused by the mismating of imperial and metric fittings or the use of non-BS male bushes.

Although some metric and imperial sizes appear to mate satisfactorily without adaptors, the differing thread form means that the earth-continuity path may not be satisfactory. The correct adaptor should always be used.

The British Standard for metric conduit fittings, BS 4568 Part 2, covers a range of male bushes which give adequate continuity when correctly installed, but some non-standard male bushes do not have sufficient width across the flats to ensure adequate continuity at all times, particularly if the entry hole is oversized. The correct BS bush should always be used.

2.20 Underground wiring in conduit

Cables installed in conduit underground must be sheathed and an earth-continuity conductor is required, in addition to the conduit, for compliance with Regulation B.125 (v) and (vi).

A twin-and-earth sheathed cable satisfies these requirements. The conduit may be heavy gauge galvanised steel, in which case it must be earthed for compliance with Regulation D.3, or it may be non-metallic, subject to compliance with Regulation B.102 as regards limits of working and ambient temperature and Regulation B.104 as regards allowance for expansion and contraction.

The note to Regulation B.125 recommends a minimum depth of 500mm. This depth must be achieved after final levelling of the ground and therefore the initial depth may well have to be greater to allow for landscaping or other changes.

2.21 Concentric cables for underground wiring

Requirements for underground wiring between buildings or between a building and the point of utilisation, as set out in Regulation B.125, apply only to consumers' wiring ond not to underground wiring on the supply side of the meter. Types of wiring used by Electricity Boards are not necessarily accepted for the conditions covered by the IEE Wiring Regulations.

Split-concentric cable to BS 4553, for example, is a non-armoured p.v.c.-sheathed cable which, according to Regulation B.125 (iv) and (vi), may only be used underground when installed in a duct, conduit or pipe, but not laid direct in the ground.

Similarly, combined neutral and earth (c.n.e.) cables are considered as non-armoured concentric cables for the purpose of the IEE Regulations and as such may not be buried direct in the ground. In addition, they should be installed in accordance with Regulations B.119–124, which deal with special requirements for earthed concentric wiring.

2.22 Temporary connections

Fatal accidents can result from carelessness in making temporary connections to a fixed installation, perhaps to maintain a supply during alterations.

The first and obvious precaution is to make sure that the supply is off before touching live conductors but it is also necessary to make sure that it is not liable to be switched on by another person while work is in progress on the connection.

Secondly, before the supply is restored it is essential to check the earthing system in both the existing installation and the temporary circuit — never take it for granted that the earthing is sound, even though the installation may have been in use for some time.

Thirdly, when taping or temporarily sealing off cable ends, particularly on multi-core sheathed cables containing an earth-continuity conductor, extra care must be taken to ensure that the cable cores and earth-continuity conductor are separated and separately sheathed.

2.23 Extension leads

Extension leads present a high degree of risk unless they are properly constructed and maintained.

Far too frequently an odd length of cable or flexible cord is pressed into service for a portable electric tool, a temporary light or even a kettle or heater, usually without an adequate socket at one end or a proper plug at the other.

An extension lead is a "temporary installation" within the meaning of Section H of the IEE Wiring Regulations and should therefore comply with all the requirements of the Regulations, for example those for mechanical protection, enclosure of terminations, adequate current rating and effective earthing. The Health and Safety at Work etc. Act has placed new legal responsibilities on employers and employees for ensuring safety at work.

An electrician may think that his specialist knowledge will enable him to avoid danger and that he has no need to take the trouble to obtain or make up a special-purpose lead, but many such leads are borrowed by other trades who do not have this knowledge with the result that the careless attitude of one man can put another man at risk.

2.24 Responsibility for making good

If the responsibilities for "making good" the damage that is inevitable on rewiring work are explicitly defined by the electrical contractor before undertaking the work, disputes on site and subsequent complaints from clients may be avoided.

It is up to the contractor to decide what is or is not included in his own responsibilities, but it should be borne in mind that the finished work must be safe within the terms of the IEE Regulations.

If making good around outlet boxes is left to other tradesmen who lack the necessary electrical skills it is inevitable that metal tools, quantities of water and moist cement will be used around the box. Socket-outlets may even be removed in order to effect a neat finish, and this could result in reversed polarity or incorrect connection of earth-continuity conductors.

One solution to this is for the electrician to make good around the outlet box so that the socket-outlet is fixed to a finished wall surface. The remainder of the wall chase can then be made good by other tradesmen without exposing them to danger or affecting the safety of the installation.

2.25 P.V.C. cables above suspended ceilings

Many new and converted buildings used for offices, schools or other purposes have false ceilings consisting of a metal framework suspended on wires and supporting ceiling tiles or panels.

This type of construction offers a useful void for the installation of building services for the supply of electricity, water, gas and heating, for example. But an electrical contractor must take particular care to comply with the IEE Wiring Regulations in such locations.

P.V.C.-insulated and sheathed cables should not be laid directly on the metal framework because of the following reasons:

1. The cables would be liable to damage by sharp edges (Regulation B.25).
2. The "inverted-T" section of the metal strips does not afford a smooth surface envisaged by Regulation B.33(v) for installation of cables without fixing. Neither can the location be deemed to be inaccessible if the ceiling tiles are removable for access to other apparatus or services.
3. Note 3 to Regulation B.53 recommends that non metal-sheathed cables should

be segregated from fixed metalwork.

4. The framework may not be designed to carry the additional weight of electric cables.

Such cables should be supported well clear of the metal framework, either by clipping to the structural ceiling or by the use of catenary wire, as appropriate to the construction of the building.

Connections to luminaires should be made by sheathed flexible cords from ceiling roses located as closely as possible above each lighting point so as to avoid fortuitous contact with the framework.

Where contact cannot be avoided, adequate protection should be provided against abrasion of the cable, bearing in mind that suspended ceilings may be subject to vibration.

2.26 Cables in roof spaces

Particular attention should be paid to the laying of cables in roof areas. In new premises, it is relatively easy to clip cables to comply with Regulation B.33 and to ensure protection against mechanical damage as required by Regulation B.25.

Roof areas of existing houses when being rewired must be treated in the same way if an access trap is provided.

In older properties, the roof space is often dirty but compliance with these regulations is still essential if there is any likelihood of anyone entering the roof space; for example a plumber attending to the water system.

Cables crossing joists should, if possible, be laid well away from access points and walkways otherwise it is recommended that the joists be drilled or mechanical protection provided as stated by Regulation B.26 for cables under floors and above ceilings.

2.27 Sheathed cables under plaster

To reduce the possibility of accidental damage to sheathed cables buried in plaster during or after construction, it is strongly recommended that the following precautions be taken:

1. On a new wall, the use of steel or plastics channelling or light-gauge conduit will provide protection against a plasterer's float; this tool is often quite sharp and can easily cut the cable sheath which would probably not be noticed at the time. Such light materials will, of course, give little, if any, protection against driven nails.

3. Cables under plaster should run vertically above or below switches, socket-outlets or other accessories because most people are aware of the likelihood of cables being in such positions.

3. If horizontal runs are essential, they should be located within 150mm of the ceiling or within a 150 to 300 mm band above floor level, where it is unlikely that wall fixings will be required.

The two latter recommendations are included in the requirements of the National House Building Council (NHBC).

Regulation B.25 requires that all conductors and cables shall be adequately protected against any risk of mechanical damage to which they may be liable in normal conditions of service. But this regulation does not mean that sheathed cables buried in plaster must be given additional mechanical protection.

If particular circumstances, however, require full protection for buried p.v.c.-sheathed cables, the use of heavy-gauge steel conduit is recommended. For domestic installations it is preferrable to avoid the risk by careful selection of the location of cable runs.

The IEE issued an interpretation in 1967 which states that cables buried in plaster should be rated for the "enclosed condition" (columns 3 and 5) in Tables 3M and 4M of the IEE Wiring Regulations.

2.28 Temporary installations

According to Regulation H.1, temporary installations should fully comply with the requirements for permanent installations except for a relaxation of Regulation B.33 in respect of supports for cables provided that no appreciable strain is imposed on terminations or joints.

In addition, Regulation H.2 requires the appointment of a competent person to take responsibility for the safety of the temporary installation; the name of such a person to be prominently displayed. Regulation H.3 states that such installations should be inspected and tested at intervals of three months or less.

Too often machines are seen working in factories with lengths of non-sheathed cables or flexible cords strung over roof trusses and pipes, for example, and connected directly to machines without local means of isolation or emergency stopping and, sometimes, with no effective earthing.

A typical explanation of this practice is that 'the machine had just been moved' and 'it was only temporary' as though this statement precluded the possibility of an accident. The exact opposite is, in fact, the case because accidents are more likely to be caused by temporary installations of the kind mentioned above than by permanent ones.

Unfortunately temporary installations have a habit of becoming permanent; likewise a death due to a temporary installation is just as permanent as death from any other cause.

3.
EARTHING and BONDING

3.1 Earthing of time switches

Failure to earth the operating mechanism of time switches is a common installation fault despite the fact that almost all time swiches, including those enclosed in all-insulated cases, are provided with an earthing terminal.

The consumer may need to make adjustments to the time switch settings, thereby making contact with metal parts such as the setting dial or levers. Unless these metal parts are separated from live parts by double insulation, connection to the earth-continuity conductor is essential for compliance with Regulation D.4.

3.2 Earthing of cable trays

Cable trays are not exempt from Regulation D.3 which requires all metalwork of wiring systems (other than current-carrying parts) to be connected to the appropriate earth-continuity conductors.

Where metal cable trays are used to support metal sheathed and/or armoured cables, they use usually adequately earthed by direct contact through metal clips to the sheath or armour which, of course, forms the earth-continuity conductor. If, however, the cable has a p.v.c. or other insulating oversheath, or if insulating clips are used, a definite earthing connection to the cable tray is necessary.

3.3 Earthing tails and sheathed wiring

An earth-continuity conductor, or "earthing tail" is not necessary between a metal box having solid lugs and the associated socket-outlet when connected to a sheathed wiring installation, because the earth-continuity conductor in the cable is connected directly into the earthing terminal of the socket-outlet. The metal box will normally be sufficiently earthed through the earthing straps and eyelets on the socket-outlet and the fixed lugs on the box.

Boxes having adjustable lugs must, however, be provided with an earthing tail between the box and the earthing terminal on the socket-outlet to ensure adequate earthing of the box, because the lug assemblies are not designed or constructed as current-carrying devices.

The necessity for this earthing tail has been demonstrated where a socket-outlet suffered overheating and burning around the fixing holes due to a line-to-earth fault onto the metal box which had not been provided with an earthing tail. Because the sliding lugs had become corroded, they presented a high resistance to the fault current flowing back to the earth-continuity conductor, and consequently the 30A ring circuit fuse did not blow.

Care must be exercised when using boxes having one fixed lug and one adjustable lug because some socket-outlets have only one earthing strap and eyelet as permitted by BS 1363; it must be ensured that the earthing eyelet is located at the fixed lug, otherwise an earthing tail must be provided.

An earthing tail is, of course, always required where a conduit or metal-sheathed cable forms the earth-continuity conductor.

3.4 Earthing tails on conduit installations

"Earthing tails" should be provided in spur connection units, flexible cord outlet plates and cooker control units with integral socket-outlets, in situations where this would be required by Regulation D.5 for socket-outlets.

This Regulation requires a connection between the earthing terminal incorporated in the associated box or enclosure where the earth-continuity conductor of the final sub-circuit is formed by conduit, trunking, duct or the metal sheath and/or armour of cables, so as to avoid relying on the plate-fixing screws as part of the earth-continuity conductor for appliance faults.

Because the final sub-circuit feeding a cooker control unit may vary in current rating, it will be necessary to refer to Table D.2M in the IEE Regulations to determine the correct size of conductor to use for the earthing tail.

This requirement for earthing tails applies not only to domestic socket-outlets and accessories but also to watertight and/or industrial type socket-outlets, to give reliable earth-continuity between the earthing terminal of the socket-outlet and the metal box or enclosure in which it is housed; indeed in some designs there would be no earth connection to the earthing contact if the earthing tail were omitted.

3.5 Loss of continuity in conduits

Loss of earth-continuity in conduit systems can be caused by a number of factors including failure to tighten male bushes, lock-nuts and cable glands correctly, failure to remove paint from a box adjacent to a conduit entry, or failure to protect against rust.

A survey of male bushes, lock-nuts and cable glands revealed that on average 10 per cent of these were slack. This is not necessarily revealed by the continuity checks, which should be carried out during the erection of a conduit system, because continuity may be good while the parts are still bright and new. Consequently continuity checks should always be supplemented by physical checks.

Use of the correct tools, including special bush spanners, and the exercise of a little more care can help to reduce the incidence of this trouble. Where a hole is too big for the bush, plain or "star" washers should be used to ensure continuity.

It is extremely difficult to obtain good continuity through the painted surface of a switch-fuse or other enclosure and it is most important to make sure that paint is removed from the surface surrounding an entry and from any conduit coupler or similar fitting before terminating conduit at a non-threaded entry..

But the paint serves more than just a decorative function — it is intended to prevent corrosion and so any "bright metal" left around the edges of a joint must be repainted, immediately after erection, to ensure that continuity is not lost through subsequent rusting.

3.6 Earth-continuity conductors in p.v.c. conduit or trunking

Where several circuits are installed in the same p.v.c. conduit or trunking, a common earth-continuity conductor will satisfy Regulation D.2 provided that its size is appropriate to the largest associated circuit conductor, in accordance with Table D.2M.

In the case of ring final sub-circuits, however, Regulations A.36 and A.52, covering ring circuits using BS 1363 and BS 196 socket-outlets respectively, state

"Except where a ring final sub-circuit is run throughout in metallic conduit, ducts or trunking, or in metal-sheathed cable, the earth-continuity conductor shall also be run in the form of a ring having both ends connected to earth at the distribution board (or its equivalent).

Where both "legs" of a ring final sub-circuit wired in 2.5 sq mm cable are enclosed in the same length of p.v.c. conduit or trunking, a separate earth-continuity conductor of 1 sq mm is required for each "leg".

3.7 Unenclosed earth-continuity conductors

Although it is permissible to connect earth-continuity conductors of 1.5 sq mm outside a joint box, subject to Regulations B.69, D.29(i) and D.29(v), the earth-continuity conductors in composite cables used for lighting circuits must not normally be connected outside the joint box.

This is because lighting final sub-circuits generally use 1.0 sq mm or 1.5 sq mm cable and the earth-continuity conductor used in composite cables of these sizes is only 1.0 sq mm. To satisfy Regulation D.28(iii), therefore, these earth-continuity conductors must be enclosed within the joint box.

It is preferable to use joint boxes large enough to permit connection of the earth-continuity conductors within the enclosure even when these conductors are of 1.5 sq mm or greater cross-section.

3.8 Bonding of metallic air ducts

To avoid a potentially dangerous situation, it is necessary to ensure that accessible ductwork and plenum chambers of electricaire, air conditioning and similar systems are effectively bonded to the earth-continuity conductor as required by Regulation D.14. The use of non-metallic seals or other components can mean that some parts of the system are not in effective electrical contact with the main body of the unit.

In one known case, relative movement within an electricaire unit resulted in a live terminal coming into contact with the plenum chamber which had not been earthed and, as a result, the ductwork throughout the house was live at mains voltage.

3.9 Bonding to gas and water services

Although Regulation D.12 states that the main bonding connections to gas or water services, should be "as near as practicable to the point of entry of those services" it is sometimes more convenient to make the bonding connections elsewhere, for example, when electricity and water services are at opposite ends of a long building.

To cover this situation, the use of certain installation metalwork as part of the bonding lead is recognised by the IEE. The metalwork so recognised is (a) pipework of other services, (b) metallic sheaths of cables other than supply service cables, and (c) metal conduit, ducts or trunking.

Attention is drawn, however, to two points. Firstly, the conductivity of any metalwork used as a bonding lead must not be less than that of the appropriate copper conductor, as given in Column 4 of Table D.2M of the Regulations. Secondly, effective bonding must not be liable to be lost due to subsequent alterations.

If pipes, cable, sheaths, etc. are used in this way, it is recommended that only main runs which are unlikely to be altered should be used and that reliance should never be placed on branch pipes or sub-circuit cables.

3.10 Clamps for bonding and earthing

A requirement for the use of clamps complying with BS 951 for bonding connections to gas and water pipes, in the first printing of the 14th Edition of the IEE Wiring Regulations, introduced an unfortunate anomaly.

BS 951 calls for a label reading "Safety Electrical Earth — Do Not Remove" to be supplied with each clamp. This is intended for use when the clamp is used with an earthing lead, enabling contractors to comply with the present Regulation D.33 which calls for such a label for earthing connections.

The use of this label on bonding connections to gas and water pipes has led Gas Board employees and others to believe that the electrical installation was earthed to the service. Consequently the reference to BS 951 clamps was removed in the 1970 amendments to the Regulations so that Regulation D.13 does not now require the label for bonding connections.

It is still necessary to use clamps for both bonding leads and earthing leads, unless soldered joints can be used, and BS 951 clamps are quite suitable for both purposes but the "Safety Electrical Earth — Do Not Remove" label should be omitted from bonding connections. In the case of earthing connections the label must be provided, even when clamps are not used.

3.11 Bonding of sinks and baths

Absence of any means of attaching a bonding lead to metal sinks and baths is a problem that concerns many contractors. Regulation D.14 calls for effective bonding in the absence of adequate continuity provided by metal-to-metal joints between taps and baths or sinks.

Two of the relevant British Standard specifications do require the provision of a means of attaching a bonding lead; these are BS 1244 dealing with metal sinks, and BS 1390 dealing with sheet steel baths. But BS 1189, which covers cast iron baths, does not have this requirement.

A number of sinks and baths are not manufactured to the British Standards and do not incorporate ready means of making the bonding connection. When these are encountered, contractors should seek to draw the attention of builders and developers to the advantages of using BS sinks and sheet steel baths.

3.12 Bonding leads under sinks

It is frequently fogotten that Regulation B.33, which requires that every cable installed in a building shall be supported so that there is no appreciable mechanical strain on any cable termination, applies equally to bonding leads under sinks.

At first-fix stage a long length of bonding lead may be left ready for connection to the water service when this is installed but when the connection is eventually made the cable is not cut to length and is left floating inside the sink cupboard where it may become displaced or damaged by buckets or other items thrown into the cupboard. The cable should be cut to length and securely supported to avoid this and, where necessary, capping should be fitted.

3.13 Use of Armouring for Earthing and Bonding

The metallic sheath and/or armouring of a cable which complies with one of the British Standards mentioned in Regulation B.1 may be used as an *earth-continuity conductor* where it forms part of the same cable as the associated circuit conductors, as permitted by Regulation D.28(iii). The sizes listed in Column 3 of Table D.2M apply only where the earth-continuity conductor does *not* form part of the same cable as the associated circuit conductors.

It is not, however, permissible to use the sheath and/or armouring is an *earthing lead* or as the *'electrical connection'* required for a protective multiple earthing (p.m.e.) supply, unless its resistance is not more than that of the equivalent copper conductor listed in Column 2 of Table D.2M For use as a *bonding lead*, the resistance of the armouring must not be more than that of the equivalent copper conductor listed in Column 4 of Table D.2M.

To determine the suitability of the armouring of cables for use as earthing and/or bonding leads, reference should be made to the tables of resistance quoted in the appropriate Standard. To assist contractors, the following table is derived from BS 6346 (Table 22) for armoured p.v.c.-insulated cables; it shows the nearest smaller copper cable size equivalent to the armour of 2-core, 3-core and 4-core cables with either copper or aluminium conductors (600/1000V grades).

nominal area of conductor	*nominal area of copper conductor equivalent to armouring*		
	2-core cable	*3-core cable*	*4-core cable*
mm^2	mm^2	mm^2	mm^2
1.5	1.5	1.5	1.5
2.5	1.5	1.5	1.5
4	2.5	2.5	4
6	2.5	4	4
10	4	4	4
16	4	4	6
25	6	6	6
35	6	6	6
50	6	6	10
70	6	10	10
95	10	10	16
120	10	16†	25*
150	16	16	25
185	16	25*	25
240	25	25	35‡
300	25	25	35
400	25	35	50

† *10 mm^2 for aluminium conductors* * *16 mm^2 for aluminium conductors*

‡ *25 mm^2 for aluminium conductors*

3.14 Earthing to water pipes

Use of water pipes as the sole earth electrode of an installation is prohibited by Regulation D.35 of the current edition of the IEE Regulations. Problems arise, however, with extensions to existing installations which were earthed to metal water pipes, as permitted in earlier editions of the Regulations.

In these circumstances, the local electricity board should be asked to provide an earthing terminal and, where they are unable to do so immediately, the consumer should be advised to install an earth-leakage circuit-breaker.

The NICEIC recognises, however, that a consumer cannot always be expected to bear this additional expense if the extension work is of a minor character and the water pipe is still effective as an earth electrode. The following procedure is therefore suggested:

1. Inspect the connection to the pipe to ensure that it is effective.
2. Test the earth-loop impedence.
3. Check with the water authority concerned that there are no plans to replace the metallic pipes by non-metallic pipes.
4. Record the deviation from Regulation D.35 on the completion certificate and advise the consumer to have the earthing tested annually, or at a shorter interval if any work has been carried out on the water supply system in the vicinity.

It is stressed that this applies only to minor extensions; new installations, rewires or major alterations must be earthed in full compliance with the IEE Regulations.

3.15 Supply authorities' earthing connections

Where cable-sheath earthing is to be provided by an electricity board, the earthing connection to their cable sheath is usually made at the time the cable is brought into the premises and the consumer's fuses and metering equipment installed. However, there are occasions when the earthing connection has not been made to the cable sheath so that there is no earthing terminal available when the contractor completes his installation work.

Under no circumstances should a contractor attempt to clamp or sweat the consumer's earthing lead direct to the supply authority's cable sheath. He should get in touch with the district office and arrange for the connection to be made by one of the board's own employees.

Similarly, the final connection to a p.m.e. earthing terminal must be made by the electricity board and not by the contractor.

3.16 Protective multiple earthing

The Area Electricity Boards are the authorities responsible for implementation of the Protective Multiple Earthing Approval issued by the Department of Energy. A Board is not permitted to connect a consumer's earthing lead to the p.m.e. earthing terminal until it is satisfied that the requirements for consumers' installations, set out in the Form of Approval, have been met.

It is always advisable to ascertain the current requirements from the appropriate District Officer of the Board concerned but, to assist contractors, the principal requirements are summarised below:

1. The only bonding required under the 1974 p.m.e. Approval is an electrical connection between the p.m.e. earthing terminal and incoming metallic services (other than telecommunications) and accessible structural metalwork.

2. The electrical connection to the incoming metallic services may, if it is more convenient, be taken from the consumer's own earthing terminal or a bar at which his own earth-continuity conductors are marshalled. Agreement of the Electricity Board should be obtained to use the consumers earthing terminal as an intermediate connection point in this way.

3. The electrical connection should be not less than 6 sq mm, associated with meter tails of 16 sq mm or less. If the meter tails are larger than 16 sq mm a larger connection must be used in accordance with Table D.2M Columns 1 and 2.

4. A common electrical connection may be used for more than one incoming service if desired. If this is done it is preferable to use an unbroken conductor or to ensure that individual lengths are permanently joined by soldering, crimping etc. to ensure that no service or accessible structural metalwork is disconnected in the event of work being carried out on one of the other services.

The 1974 p.m.e. Approval uses the term "electrical connection" to avoid confusion with the bonding leads referred to in Regulations D.10-14. These Regulations must be observed. The electrical connection to incoming metallic services and accessible structural metalwork will normally satisfy the equivalent requirements of Regulations D.10-14 but additional local bonding may be required to satisfy the full requirements of Regulation D.14.

3.17 Bonding on extensions

When making major alterations or extensions to existing installations, it is important to check whether bonding exists to satisfy Regulations D.10 and D.14. If not, provision must be made for the necessary bonding to be undertaken in conjunction with the new work.

With small alterations, such as an additional outlet point connected to an existing final sub-circuit, it may be necessary to install local bonds to satisfy Regulation D.14 because many installations originally completed before 1966 have no bonding of any kind.

This situation is likely to arise when an additional point is installed for immersion heaters, gas or oil-fired boilers or when a new socket outlet is positioned near a metal sink.

Where there are reliable metal-to-metal joints in the pipework of other services, main bonds to the requirements of Regulation D.10 (minimum size 6 sq mm) will also satisfy Regulation D.14.

3.18 Terminating armoured cables

Steel-wire-armoured p.v.c.-insulated cables installed in conjunction with Class II (all-insulated or double-insulated) apparatus or switchgear are sometimes terminated without using glands and the strands of the armour taken directly to the equipment's earthing teminal.

Apart from the loss of mechanical protection for the cable, this practice is undesirable because terminals will often not accept the full number of strands with the result that a proportion are cut away leading to a contravention of Regulations B.61 and D.28(iii).

In view of this, the correct gland having an earth bond attachment should always be employed for terminating armoured cables at Class II equipment.

Similarly for Class I (metalclad) equipment, a type 'C' gland to BS6121 will ensure continuity between the armouring of a p.v.c./s.w.a./p.v.c. cable and the gland itself and also to the metal enclosure of the equipment.

Where such glands are used with thin steel enclosures, it is not advisable to rely on clamping by a back-nut for continuity between the gland and the enclosure; for such conditions an earth bond attachment should also be used.

The use of an earth clamp, whether to BS951 or not, direct on the armour is not satisfactory because it necessitates stripping back the outer p.v.c. sheath thus destroying the protection against corrosion.

Additionally, it is difficult to make a secure connection because the wires forming the armour are compressed into the p.v.c. bedding so that the effectiveness of the clamp will reduce with time.

3.19 Earthing of conduit

The need to ensure that conduit terminations are fully tightened when they form part of the earth-continuity path is well understood.

Equal attention, however, should be paid to this point when the conduit is not used as the earth-continuity conductor when, for example, a separate earth wire has been run inside the conduit.

It is still necessary to ensure that the conduit is correctly earthed in accordance with Regulations B.98 and D.3 and a check should always be made to make certain that all conduit terminations are mechanically and electrically sound.

3.20 Fire alarm call points

Plastics break-glass call points are now being supplied by some manufacturers.

Where these are used with metallic wiring systems (conduit or metal-sheathed cables), installers should make sure that all necessary earth connections have been made to maintain earth-continuity across the box.

Tests of earth-continuity impedance, in accordance with Regulation E.3, should always be undertaken before commissioning every installation.

3.21 Responsibility for Earthing

The average consumer is not aware that he is responsible for ensuring that his electrical installation is correctly protected against earth-leakage.

In practice, the electrical contractor assumes this responsibility on behalf of the consumer, by ensuring that the installation has been properly earthed in accordance with the IEE Wiring Regulations before he issues a Completion Certificate.

Contractors are urged, therefore, to ascertain at an early stage whether an earthing terminal will be made available by the Area Electricity Board. If not, it is essential to determine visually and by electrical tests the type of earthing that will be needed for each installation.

Electricity Boards are not bound to provide earthing terminals for use by consumers, but they will normally do so for new premises and, on request, for rewires, where the supply distribution system permits this to be done.

The following procedure should be adopted:

1. Check whether or not there is a suitable earthing terminal already available at the service position.
2. If not, ascertain whether the Electricity Board is prepared to provide one.

3. If an earthing terminal is available, check with an earth-loop impedance tester to see if it is effective.
4. If the answers to (2) and (3) are no, then decide whether to provide (i) an earth electrode to allow the use of direct earthing (Regulation D.22) or (ii) an earth electrode and an earth-leakage circuit-breaker — preferably current-operated — in accordance with Regulation D.23.
5. Finally, before the installation is handed over for use, re-check, by testing, that the means of earthing is effective.

4. ISOLATION and PROTECTION

4.1 Main isolating switches

A number of factors are worth considering when selecting suitable means of isolation for supplies to domestic and small commercial premises.

Except where earth-leakage circuit-breakers are needed to ensure compliance with Section D of the Regulations, consumer units are often used.

The double-pole main switch with shrouded incoming terminals provides a convenient and compact means of isolating the whole installation and, because of its simplicity, is unlikely to need attention during the life of the installation.

A circuit-breaker, with or without excess-current and/or earth-leakage protection, is an acceptable alternative but is a little more complicated. If it needs attention it can only be made dead by removal of the supply authority's service fuses.

The current rating of an isolating switch for a distribution board should not be less than the protection provided by the fuses or circuit-breaker controlling the circuit. In small installations, where this protection is provided by the supply authority's service fuse, this can present problems if a consumer unit incorporating a main switch rated 60A, or even 45A, is used to control a supply derived from service fuses rated at, say, 100A. The main switch and meter tails could be overloaded if extensions are carred out without due consideration of this point.

In designing an installation it should always be borne in mind that one of the principal purposes of the main isolating switch is to provide means of switching off the supply in an emergency. If aesthetic considerations banish it to some inconvenient place under stairs or in a garage, it can hardly be considered suitably located for this purpose.

4.2 Isolation in large buildings

While a main isolator is necessary for the incoming supply to the consumers' installation in a large building such as a block of offices, consideration of convenience and safety demand additional isolation of supplies to particular parts of such buildings.

In tall buildings it is wise to provide means of isolation at each floor to avoid the inconvenience of cutting off the supply to the rising main. Similar precautions can be taken in long buildings.

A single tapping from a metered rising main can be made through an adjacent switch-fuse. For a multiple tapping, e.g. to serve a number of separate areas, it may be more convenient to provide an isolator from which a supply is taken to a distribution board. Alternatively, cartridge fuses could be used in the same way as service fuses, but the attendance of an electrician would be required before disconnection could be effected whereas switches or fuse-switches would be available for general use in emergencies.

Tappings from rising mains will only be protected against short circuit by the relatively heavy-current protection at the bottom of the building and it is therefore essential to comply with Regulation A.11 or A.12 as applicable. Whatever means of isolation is adopted, the live terminals must be screened to allow safe access to the "dead" side.

Where unmetered rising mains are to be installed, the proposed arrangements should be subject to discussion with the supply authority.

4.3 Current rating of distribution fuseboards

When a distribution fuseboard is intended to supply only fixed loads to which a diversity factor cannot be applied, such as off-peak heating systems, it is important to check with the manufacturer that the board is adequately rated for this purpose, in accordance with Regulation C.3.

The permissible total loading of fuseboards to BS 214 is determined by the application of a diversity factor ranging from 100% down to 66% depending on the number of fuseways and their nominal current rating. The maximum size of conductor which can be accommodated in the incoming terminals is determined in the same manner.

Where all fuseways of a distribution board are to be used for constant loads, it will often be necessary to install a board with a higher nominal rating to ensure that the necessary incoming cables can be accommodated and that the board is not overloaded. The fuses used should, of course, be appropriate to the cable sizes for the individual sub-circuits.

In the case of consumer units to BS 1454, the total current rating is determined by the nominal current rating of the main switch in the unit. The total continuous load must not exceed this value.

4.4 Check fuses before use

Installation equipment should always be checked at the time when it is installed to ensure that it is complete and suitable for the application. This is especially true of equipment providing excess-current protection.

Cartridge fuses, for example, should be checked to ensure that the screws securing the fuse links to the carriers have been properly tightened, otherwise overheating and damage may occur.

Similarly, rewireable fuses which have been supplied in pre-wired form should be checked to ensure that each fuse rating is appropriate to the sub-circuit protected; unless instructed otherwise, manufacturers wire all fuses for the maximum rating of the fuse carrier.

4.5 Class of excess-current protection

The concept of "close" and "coarse" excess-current protection was introduced with the current (14th) Edition of the IEE Wiring Regulations. The terms are defined in the definitions to the Regulations and explained in more detail, with examples, in the preface to the tables of current rating and voltage drop.

Close excess-current protection is protection which *will* operate within four hours at 1.5 times the designed load current of the circuit which it protects. Coarse excess-current protection *will not* operate within four hours at 1.5 times the designed load current.

The class of excess-current protection afforded by different forms of protective device is shown in the following table.

Protective device	*Fusing or tripping factor*	*Class of protection*
Cartridge fuses to BS 88 Part 2		
Class P	not over 1.25	close
Class Q1	not over 1.5	close
Motor-circuit	over 1.5	coarse
Cartridge fuses to BS 1361	not over 1.5	close
Semi-enclosed (rewireable) fuses to BS 3036	not over 2.0	coarse
Miniature and moulded-case circuit-breakers to BS 3871	1.25–1.5	close
Adjustable circuit-breakers set to operate at:		
(i) not more than 1.5 times designed load current of the circuit	not over 1.5	close
(ii) more than 1.5 times the designed load current of the circuit	over 1.5	coarse

4.6 Use of miniature circuit-breakers

The following points should be considered when designing an installation which incorporates miniature circuit-breakers.

Performance characteristics for m.c.b.s. manufactured to BS 3871 Part 1 are specified for an ambient temperature not exceeding 30degC. If this temperature is exceeded, the tripping current and fault-breaking capacity may be reduced. If possible, the m.c.b. board should be located where this temperature will not be exceeded, but if this is unavoidable advice should be sought from the manufacturer.

On large commercial or industrial installations, especially when the m.c.b. board is situated near the main busbar, careful consideration should be given to the relationship between the short-circuit duty of the m.c.b. and the prospective fault current. BS 3871 Part 1 provides for a range of short-circuit duty categories M1 to M9 which correspond to prospective fault currents from 1kA to 9kA. Where the prospective fault current exceeds the available short-circuit duty of m.c.b.s, it may be possible to use moulded-case circuit-breakers (m.c.c.b.s) to BS 3871 Part 2. This provides for short-circuit duties ranging from 10kA to 43kA.

4.7 Shock protection by e.l.c.b.

It is important to distinguish clearly between direct and indirect contact with live parts when considering the degree of protection against shock provided by earth-leakage circuit-breakers.

The only accepted method of preventing shock risks due to direct contact with live metal is by enclosure of all live conducting parts in such a way that they are not accessible to touch when the apparatus is installed and in service.

Protection against indirect contact (i.e. when metal parts become alive due to failure of insulation or other causes) is afforded by earthing such parts, or by separating them from live parts by means of double insulation, or similar methods.

Where earthing is adopted for protection against indirect contact (i.e. all normal installations), Regulation D.20 requires that the supply should be disconnected before damage can occur in the event of an earth fault. This is to be achieved by the use of excess-current protective devices in accordance with Regulation D.22 or, where these conditions cannot be satisfied, by one or more earth-leakage circuit-breakers in accordance with Regulation D.23.

It is sometimes suggested, incorrectly, that a high-sensitivity current-operated e.l.c.b. (rated operating current 30mA or less) will also provide protection against direct contact, since fatalities do not normally occur at currents of 30mA or less. It must be remembered, however, that an e.l.c.b. is not a current limiting device and so the full prospective fault current will flow, subject to circuit resistance, until the e.l.c.b. operates. Under favourable conditions the operating time may be sufficiently short to prevent a fatal shock, but it cannot be relied upon to prevent a shock sensation that may be sufficient to cause secondary injury due, for example, to the victim falling off a ladder.

Nevertheless, the high-sensitivity current-operated e.l.c.b. does provide useful additional protection where failure of other protective measures, or carelessness on the part of users, could create a shock risk.

4.8 Testing of earth-leakage circuit-breakers

Every earth-leakage circuit-breaker is provided with a test button which should be used regularly by the consumer. If this check is omitted, the e.l.c.b. is liable to stick in the "on" position.

The test button only checks the mechanical and electrical operation of the e.l.c.b. It does *not* check that the unit will operate under fault conditions because the earth-loop path is not included in the test circuit.

A test for effectiveneses of e.l.c.b.s is required by Regulation E.5 and is described in item (3) of Appendix 6. A voltage of not more than 45V is applied from a double-wound transformer which should cause immediate tripping of the e.l.c.b.

4.9 Voltage-operated e.l.c.b. snags

The voltage-operated earth-leakage circuit-breaker suffers a number of drawbacks outlined below and the current-operated device is to be preferred wherever practicable.

It is difficult to achieve discrimination between several voltage-operated e.l.c.b.s on the same installation, or even on different installations. An e.l.c.b. may trip as a result of a fault imported from another consumer's installation.

Parallel earth paths through extraneous earthed metal such as pipes or structural metalwork of a building can have an impedance low enough to shunt the trip coil of the voltage-operated e.l.c.b. and so prevent operation; and yet the same earth path may not be of sufficiently low impedance for the excess-current protection to operate.

The voltage-operated e.l.c.b. will not operate without an effective earth connection because sufficient current must flow from the protected metalwork, through the trip coil and the earthing lead to the earth electrode. The current-operated e.l.c.b., on the other hand, is not affected by parallel earth paths because its operation is due to the imbalance in a differential-current transformer when a leakage current flows to earth instead of returning through the neutral conductor.

4.10 Voltage-operated e.l.c.b.s. for on/off peak installations

Problems can arise when more than one voltage-operated earth-leakage circuit-breaker is used to protect parts of the same installation, for example for normal and off-peak supplies.

Where two voltage-operated e.l.c.b.s are used to give the separate control required by Regulation A.1, the "frame" side of both circuits will be common if cross-bonding has been carried out to comply with Regulations D.10 to D.14. Hence both e.l.c.b.s are likely to operate in the event of a fault on either part of the installation.

Another problem stems from the incorporation in some voltage-operated e.l.c.b.s of a short-circuiting switch across the trip coil to prevent burning out of the coil when the circuit-breaker is in the "off" position. Unfortunately, due to the difficulty of ensuring that the resistance areas of the two earth electrodes do not overlap, it is possible for one e.l.c.b., if it is switched off, to shunt the trip coil of the other so as to prevent it operating if a fault should occur.

One solution to these problems is to use one double-pole voltage-operated e.l.c.b. arranged so that each pole isolates the live conductor of one supply. Although this approach does not comply strictly with Regulations A.1-4 and should be recorded as a deviation, it may be regarded as providing a degree of safety from fire and shock not less than that assured by compliance with the Regulations.

If a 250V rated e.l.c.b. is installed, this arrangement can, of course, only be used where the normal and off-peak installations are connected to the same phase of supply. A 500V rated unit must be used if two phases are present.

4.11 Electro-damp proofing and e.l.c.b. protection

Voltage-operated earth-leakage circuit-breakers can be rendered inoperative by a form of damp proofing which is used by some local authorities and developers in house modernisation schemes.

The system employs a copper conductor fixed around the building at, or near, ground level and connected to earth electrodes driven into the ground. Such a system could easily short-circuit the operating coil of a voltage-operated e.l.c.b. since the conductors of the damp proofing system, the incoming gas and water service pipes, and the earth electrode of the e.l.c.b. may well be within the resistance area of one another.

Current-operated e.l.c.b.s, which do not suffer from this disadvantage, should preferably be used on rewire programmes where such a system of damp proofing may be used.

4.12 Earthing connections for e.l.c.b.s.

Certain requirements should be met by earthing leads and earth electrodes with both current-operated and voltage-operated earth-leakage circuit-breakers. These are summarised below whilst special requirements for each type of e.l.c.b. are covered separately.

The earthing lead for an installation incorporating an e.l.c.b. must be adequately protected from mechanical damage (Regulation B.25), not less than 2.5 sq mm (Regulation D.31) and protected by green insulation (Regulations B.54(i), D.29(i) and D.32). When the earthing lead is buried in the ground it should be enclosed in conduit or armoured to provide mechanical protection.

The connection to the earth electrode must be readily accessible, soundly made by soldered joints or clamps, and provided with a label reading "Safety Electrical Earth — Do Not Remove" (Regulation D.33). The earth electrode can be protected and yet made accessible by constructing a brickwork surround with a cover tile on top or using a cast iron inspection cover of the type used for water stop-cocks.

The earth electrode should be as near as practicable to the consumer's earthing terminal and must satisfy the requirements of Regulations D.24 (for current-operated e.l.c.b.s) or D.25 (for voltage-operated e.l.c.b.s) as appropriate.

Even on extension work, where the extension relies on an existing e.l.c.b., these requirements are related to the new work within the meaning of Regulation E.1. If, for example, the existing earthing lead disappears into the ground, it must be exposed and traced to the earth electrode, the integrity of the connection must be checked and mechanical protection must be provided for both the lead and the electrode.

4.13 Earth electrodes for voltage-operated e.l.c.b.s

The importance of effective earth electrodes for voltage-operated earth-leakage circuit-breakers cannot be over emphasised. Fatalities often occur as a result of broken earth connections.

In addition to the requirements detailed above for all types of e.l.c.b., the following points should be borne particularly in mind with the voltage-operated types.

A "suitable" earth rod must be used, preferably a commercially manufactured design. A piece of black enamelled conduit, for example, is not a suitable electrode.

Care must be taken to install the earth rod outside the resistance area of any parallel earth path such as gas, water or oil pipes or the structural metalwork of the building. If this is not observed, the operating coil of the e.l.c.b. will be partially short-circuited and the device could fail to operate within the 40V limit. A separation of 2.5 – 3m between the earth electrode and other earthed metalwork is usually considered adequate.

Voltage-operated e.l.c.b.s which comply with BS 842 are designed to operate at 24V with an earth electrode resistance of 200ohm and at 40V with a resistance of 500ohm. Although soil resistivity varies around the country, the higher value should not be difficult to achieve with a suitable earth rod.

4.14 Earth electrodes for current-operated e.l.c.b.s

An earth electrode is necessary, even where an installation is protected by a current-operated earth-leakage circuit-breaker, to make certain of a definite earth-fault-loop path. If this is not provided, metalwork could remain alive under earth-fault conditions until the earthing circuit is completed, possibly through the body of a person making simultaneous contact with other earthed metalwork.

Use of metal water pipe as the earth electrode for a current-operated e.l.c.b. is no longer acceptable. The new Regulation D.35, introduced with the April 1976 amendments to the IEE Regulations, states explicitly "The metalwork of public gas and water services shall not be used as the sole earth electrode of the installation." An independent earth electrode must be used or the connection made to the supply authority's cable sheath, if permitted.

Other requirements for an earth electrode for an e.l.c.b. are listed under "Earthing connections for e.l.c.b.s" above.

4.15 Selection of earth-leakage circuit-breakers

Now that current-operated earth-leakage circuit-breakers (e.l.c.b.s) are available with operating currents of 30mA and less, it is possible to select a suitable device for almost any location so it should seldom be necessary to use a voltage-operated unit which lacks discrimination and is susceptible to non-operation due to parallel earth paths.

For a current-operated e.l.c.b., the table shows the relationship between the operating current and the maximum permitted earth-loop impedance for compliance with Regulation D.24.

Operating current mA	Maximum earth-loop impedance ohms
1,000	40
500	80
300	133
100	400
50	800
30	1,333
25	1,600

In practice, the higher values of earth-loop impedance shown in the table (say, 400 ohm and over) would be exceptional and there would be good reason to doubt the reliability of an earthing system dependent on such values: steps should be taken to improve the earth-loop path or to request the provision of a protective multiple earthing (p.m.e.) terminal.

While the use of high-sensitivity current-operated e.l.c.b.s is not precluded for the protection of complete installations, the main application for such units is to give additional shock or fire risk protection for installations which already have earthing to satisfy Regulation D.22 but where hazardous conditions exist so as to justify their additional cost.

For example, the Code of Practice for Beer Dispense in Licensed Premises recommends the use of 30mA current-operated e.l.c.b.s protect circuits in cellars and elsewhere.

Similarly, the recommended method of providing a p.m.e. supply to touring caravan sites includes the provision of sensitive e.l.c.b.s for groups of up to six caravans.

4.16 Luminaires and Regulation A.60

Since Regulation A.60 was amended to permit the switch required by Regulation A.58 to be incorporated in a luminaire, some confusion has arisen regarding the type of luminaire which is covered by this regulation.

Regulaton A.60 contains two conditions. First, it must be possible to dismantle the luminaire for maintenance (lamp replacement or normal cleaning, for example) without exposing live parts when the switch is "off". Second, any internal wiring which remains live when the switch is "off" must be segregated from all other internal wiring by screens of earthed metal or suitable barriers of insulating material.

Unless both these conditions are met, it is not permitted to rely only on the switch in the luminaire and a separate switch must, therefore, be provided in accordance with Regulation A.59.

This requirement will apply generally to wall-bracket type luminaires and certain types of strip light in which the wiring may not be segregated.

5.
LIGHTING

5.1 "Home Office Skirts"

Bayonet cap lampholders should be fitted with BS 52 protective shields — commonly known as "Home Office skirts" — when they are installed within 2.5m of baths or shower cubicles (Regulation D.16) or in damp situations or situations where they can be readily touched by a person in contact with, or standing on, earthed metal (Regulation C.23).

The requirement applies to both pendant and batten lampholders. Typical situations where the requirement of C.23 applies would include kitchens where the lampholder is mounted above the sink or cooker, toilets, cloakrooms and in roof spaces adjacent to pipework or water storage tanks. It may also apply to living rooms and other areas, due to the presence of central heating radiators and other fixed metalwork.

The object of this requirement is to make it impossible to touch the metal lamp cap, when cleaning or replacing the lamp, in case a defective lamp should cause the cap to become live. The extended shade ring fitted to many lampholders does not satisfy this requirement.

5.2 Bayonet cap lampholders

Moulded, insulated bayonet-type lampholders are required, by Regulation C.24, to bear the marking "BS 52A" or, if they are likely to attain a temperature in excess of 135degC, "BS 52H". This signifies compliance with the relevant sections of BS 52.

A 100W lamp used in the cap-up position is likely to create temperatures in excess of 135degC and so the heat-resisting "BS 52H" lampholder should be installed in any location where lamps of this rating, or higher, are likely to be used.

The use of clearly-marked BS 52 lampholders is important and will avoid the danger which could arise from some inferior products which are not even suitable for lamps of 40W or less.

5.3 Suspension of lighting trunking

Methods of suspending lighting trunking sometimes leave much to be desired when the weight of the luminaires, cables and the trunking itself are all taken into consideration.

Lightweight close-link chain is inadequate for this purpose and has the added disadvantage that it permits undue sideways deflection of a long run of trunking. Suspension should preferably be by conduit or rod, but if chain suspension is essential, only welded chain should be used.

It is also important not to exceed the spacing recommended by the manufacturer for suspension points, taking into account the number and weight of luminaires to be installed.

5.4 Luminaire track and phase of supply

Wherever possible, luminaire track systems should be connected to the same phase as the socket-outlets in the same vicinity for compliance with Regulation A.20.

Where the use of two or three phases is unavoidable, track sections on different phases must be separated by at least 2m from each other and from socket-outlets where there is a phase difference.

A luminaire track system complying with BS 4533 is regarded as one point, for the purpose of Regulation A.26, provided that individual luminaires are suitably protected against excess current. This means that the sub-circuit fuse may exceed 15A, depending on the current rating of the track.

5.5 Final connections to luminaire track

Final connections to luminaire track systems are subject to the requirement of Regulation B.69 which calls for enclosure of cable cores.

The relevant British Standard, BS 4533 Part 2, Section 2.6, requires that cable entries should allow the introduction of the conduit or the protective covering of the cable or flexible cord so that the cores are completely protected. Contractors should ensure that the correct components are ordered for a luminaire track system to avoid any departure from Regulation B.69.

5.6 Cable termination at enclosed luminaires

Electricians sometimes loop p.v.c.-insulated cables into enclosed tungsten luminaires without due regard to temperature conditions although it is well known that high temperatures can arise in the fittings, especially multi-lamp types and those with high wattage lamps.

Failure to provide adequate protection of the tails by heat-resisting sleeving results in the insulation melting at the crutch, where the cores enter the cable sheath. Use of p.v.c.-insulated connectors which have subsequently melted leaves bare live conductors within the luminaire.

Where connection is to be made from a sheathed wiring system to an enclosed luminaire in which the temperature may exceed 70degC, a joint box should be fixed in the roof void or floor space adjacent to the fitting and the final connection should be made in heat-resisting cable (See Table B.3M).

Alternatively, where a joint box cannot be used because of inaccessibility, a metal conduit box complete with lid should be fitted flush with the ceiling and the connections made within this box using porcelain-insulated or other heat-resistant connectors. This method is also appropriate for a conduit installation.

A further temperature reduction can be achieved by fitting an extension ring to the conduit box, thereby ensuring free circulation of air around the luminaire and its connections.

5.7 High temperature in fluorescent luminaires

It is not always realised that high temperatures can arise in fluorescent luminaires, making it necessary to pay close attention to wiring within the fitting, and to observe any recommendations by the manufacturer regarding the use of heat-resisting cables.

The control gear in some fluorescent luminaires may operate at temperatures well in excess of the 70degC permitted for general-purpose p.v.c. insulated cables. The cables employed by the luminaire manufacturer for internal wiring should be suitable for the design temperature but if other cables are to be routed adjacent to the control gear, the cores must either be sleeved individually with heat-resisting insulation such as 85degC or 150degC rubber, or glass fibre (See Regulation B.36).

These measures may be avoided by using the entry immediately adjacent to the terminals of the luminaire. In some cases this will be off-centre and it is therefore desirable to obtain advance information of the type of luminaire to be used so that conduit boxes etc. can be correctly positioned at the first-fix stage.

5.8 Luminaires and p.v.c. conduit boxes

Where a p.v.c. conduit box is used to support a luminaire, or is in contact with a luminaire, the mass suspended from the box must not exceed 3kg and the temperature of the box must not be permitted to exceed 60degC, to satisfy Regulation B.70.

Even where two of these boxes are used to support a fluorescent luminaire, the mass on each box is likely to exceed 3kg. Furthermore, BS 4533 permits the temperature of the supporting surfaces of a luminaire to rise to 80degC under normal operating conditions; under "abnormal conditions" such as failure of one tube of a twin-tube luminaire the surface temperature is permitted to rise to 180degC. It does not follow that all luminaires will attain the first of these temperatures but, in the absence of definite information from the manufacturer, it must be assumed that they will do so.

To use normal p.v.c. conduit boxes under these circumstances would clearly be a deviation from B.70. There are, however, boxes covered by Assessment of New Techniques Scheme (ANTS) certificates which are suitable for temperatures up to 100degC and suspended masses in excess of 3kg.

5.9 Control gear separated from lamps

Control gear which is separated from the associated lamps, as in showcase or canopy lighting in shops, must be enclosed to satisfy Regulation G.5.

Control gear mounted on trays with covers is available for this purpose. Separate chokes with exposed terminals, intended for building into luminaires, are not suitable unless provision is made for enclosure of the connections as required by Regulation B.69. In addition, all exposed metalwork of the equipment must be connected to the earth-continuity conductor.

It is also important to ensure that cables and flexible cords used for connection between the control gear and lampholders are sheathed or enclosed to comply with Regulations B.28, B.69 and B.130.

5.10 Three-phase supplies for lighting

High levels of illumination in factories, supermarkets and other locations frequently necessitate provision of a three-phase supply for lighting. When possible, the whole of such a load should be controlled by a contactor on the supply side of the lighting distribution board, without the use of individual switches.

On small installations, or where local switches are essential, there is a tendency to use multi-gang grid switches of 250V rating. As it is virtually impossible to segregate phases within a single multi-gang switch enclosure, as required to comply with Regulation A.18, it is recommended that separate switch boxes be used for each phase. Alternatively, industrial type switches are available in cast iron boxes which incorporate barriers and separate compartments for each phase, or purpose-made boxes can be fabricated by a specialist in sheet metalwork.

Care must be taken that each group of switch wires is contained within the same conduit and enclosure as the associated phase feed wire, as required by Regulation B.100.

In addition, a separate neutral conductor must always be provided for each group of luminaires supplied from a separate fuse-way (Regulation A.24). Looping of a single neutral conductor to all fittings would be a deviation from Regulation A.24 and, in the case of fluorescent lighting, would lead to difficulties with third harmonic currents which do not balance to zero. For the same reason, the cross-sectional area of the neutral conductor feeding a three-phase distribution board for fluorescent lighting must not be reduced below that of the live conductors.

5.11 Overloading of fluorescent lighting circuits

Failure to make adequate allowance for increase in current due to power factor, choke loss and harmonics in fluorescent lighting installations will result in overheating of supply cables.

Regulation G.3 recommends that in the absence of more exact information from the luminaire manufacturer, the rated lamp watts should be multiplied by a factor of 1.8 to give the volt-ampere demand. Using this rating factor the maximum number of single-tube luminaires that could be supplied from one 5A or 15A circuit would be given by the following table:

Tube rating	*No. of single-tube luminaires*	
	5A circuit	*15A circuit*
40W	16	50
65W	10	30
80W	8	25
125W	5	16

To take full advantage of these figures, switches must be rated for inductive loads in accordance with Clause 23 of BS 3676. Switches intended only for non-inductive loads must be de-rated by a factor of 0.5 if used to control fluorescent lighting. Additionally, in selecting cable sizes, account must be taken of volt drop, ambient temperature, grouping and class of excess-current protection.

In some cases, use of the 1.8 rating factor may be unduly restrictive, for example on circuits where the power factor is corrected to 0.95 or better, and reference should be made to manufacturers' literature to determine the actual operating current. Other circuits, however, with a low leading power factor of about 0.65 may result in a higher current than indicated by the 1.8 rating factor.

5.12 Control of discharge lighting circuits

To comply with Regulation G.12, a fireman's emergency switch should be installed for both exterior and interior discharge lighting installations operating at more than 600V to earth. This requirement does not apply to portable interior luminaires nor to signs of 100W rating or less and fed from an adjacent socket-outlet.

In addition to this switch, Regulation G.11 requires; inside the building, a means of isolation from all poles of the supply by one of the following methods:

1. An interlock on a self-contained luminaire, so arranged that the supply is automatically disconnected before access can be gained to live parts.
2. A conveniently placed socket-outlet or other effective means of isolation.
3. A switch having a removable handle or lockable switch or distribution board. Where more than one such item is installed, all keys or removable handles shall be non-interchangeable.

Methods (1), (2) and the lockable distribution board in method (3) are additional to the switch normally used to control the luminaire or circuit.

6. SPECIAL ENVIRONMENTS

6.1 Sauna baths

A sauna bath should be treated in the same way as any room containing a fixed bath or shower, as far as protection against earth-leakage currents is concerned, and is therefore subject to the additional requirements set out in Regulations D.15–19 and A.61. The position has been set out in a statement by the Institution of Electrical Engineers reproduced below.

"Unconventional means of bathing were not contemplated when Regulations D.15–19 were devised. It is considered, however, that these regulations should, by analogy, be applied where a person occupying a sauna bath is in a condition comparable to that of a person using a normal domestic bath, i.e. unprotected by clothing, possibly in a moist condition, and in a situation where earthed metal, such as the metal frame of the heater or tray of pebbles, is accessible.

"In such circumstances, it is undesirable that controls, which may be of material susceptible to the formation of a surface film of condensation, should be located where they may be handled by a person free to move about the room. Regulations D.17 and A.61 are therefore applicable and all controls should preferably be sited outside and immediately adjacent to the normal entrance to the room."

6.2 Socket-outlets near swimming pools

While some of the IEE Regulations' requirements for installations in bathrooms are readily applied to indoor and outdoor domestic swimming pools, for example Regulation D.16 (for lampholders) and D.17 (for switches and heating appliances), strict application of Regulation D.18 would prohibit installation of socket-outlets adjacent to such pools. This would be likely to result in the undesirable use of excessive lengths of flexible cord for portable pool-cleaning equipment and other appliances.

Where socket-outlets are installed adjacent to swimming pools it is recommended, firstly, that they be located not only out of reach of a person using the pool but far enough away to avoid splashing and, secondly, that they be supplied through a current-operated earth-leakage circuit-breaker having a rated operating current not more than 30mA. A lower operating current is preferable.

6.3 Kitchens used as bathrooms

Some older dwellings have a fixed bath installed in the room normally used as a kitchen. Although provision of socket-outlets in such rooms constitutes a deviation from Regulation D.18, a complete prohibition on the installation of socket-outlets would deny the occupant the use of electrical equipment in the kitchen area when the bath was not in use.

Where an electrical contractor is asked to install socket-outlets in such rooms, he should first establish that the room was designed to be used as a kitchen and is still normally used for purposes other than a bathroom. If he is satisfied that this is the case, the following procedure is recommended.

1. Each socket-outlet should be situated so as to be inaccessible to a person using the bath and should have a notice fixed to the wall, adjacent to the socket-outlet, reading "IMPORTANT. Before using the bath, remove the plug from this socket".
2. A cooker control unit which incorporates a socket-outlet should also be

situated so as to be inaccessible to a person using the bath, and should have a notice fixed to the unit, or on the wall adjacent to it, reading "IMPORTANT. Before using the bath, switch off the control switch and remove the plug from the socket".

Note: Both of these labels should have red lettering and be fixed so that they remain visible at all times.

3. The completion certificate, required by Regulation E.11 to be handed to the person ordering the work, must include details of the departure from the Regulations together with reference to the special measures taken under items 1 and 2 above.

4. Socket-outlets should not be installed solely for space heating appliances; if electrical space heating is required, the use of suitable fixed appliances, installed in accordance with Regulation D.17, is recommended.

6.4 Folding shower cubicles

To take account of the use of folding shower units, Regulation D.18 was amended in 1974 to read "Where shower cubicles are located in rooms other than bathrooms, any socket-outlets shall be situated at least 2.5m from the shower cubicle."

Where such a cubicle is installed, switches and other electrical controls, and appliances with touchable heating elements, must be located out of reach of a person using the shower (Regulation D.17), lampholders must be shrouded or located at least 2.5m from the shower cubicle (Regulation D.16) and socket-outlets must also be at least 2.5m from the cubicle although shaver units complying with BS 3052 may be located within this distance (Regulation D.18).

If, at the time of wiring, a contractor is aware of the intention to install a shower unit in a bedroom, the above Regulations should be applied in full. If, however, the shower unit is installed later, no responsibility for failure to comply with Regulations D.16–18 can devolve on the original electrical contractor.

6.5 Bathroom cabinets

Some bathroom cabinets sold through department stores incorporate socket-outlets, intended for electric shavers, which do not comply with the requirements of BS 3052 for bathroom use or incorporate the current limiting device required by BS 4573 for rooms other than bathrooms — See Regulation F.7(ii).

If an electrical contractor is called in to connect such a cabinet, it is important that he check whether it is suitable for bathroom use and advise the customer accordingly.

6.6 Positioning of shaver supply units

Shaver supply units are not splash-proof and so, although Regulation D.19 permits the use of BS 3052 shaver supply units in bathrooms, this does not mean they may be positioned anywhere in the room.

Normal switch height will usually safeguard against splashing from the wash basin. The unit should also be positioned well away from the bath or shower. The possibility that shower fitments may be added after the electrical installation has been completed, should also be borne in mind.

6.7 Split-level cooker installation

Split-level cookers, where the oven and hob are separate units, sometimes make it difficult to find a suitable position for a single control switch for the whole equipment because of the requirement in Regulation A.29 that the switch must be located within 2m of both appliances.

Where this cannot be achieved, separate switches should be provided for each appliance although a single final sub-circuit is acceptable where the total rating does not exceed 30A when assessed in accordance with Regulation A.27.

Although not specifically stated in the Regulation, the switch or switches should be readily accessible at all times — never placed below the working surface or in a cupboard.

Where a single switch is used, it should be noted that the cable size for connecting each of the cooking appliances cannot be reduced below the size required between the distribution board and the switch unless additional local fuse protection is provided.

6.8 Damp situations on farm premises

Farm buildings which have slatted sides for ventilation come within the category of "damp situations" under which the requirements of Regulation C.6 must be applied.

The slats admit rain and snow so that equipment installed below them is subject to conditions similar to those outside. Deterioration of ordinary apparatus is rapid under such conditions. It is essential, therefore, to use damp-and-dust-proof apparatus with appropriate cable or conduit entries as indicated in Regulation C.6.

6.9 Garages and service stations

The principal requirements for electrical installations in garages and service stations where petrol-engined vehicles are stored or repaired are set out below. No special requirements apply for premises used solely for the repair of diesel-engined vehicles, nor to individual domestic garages.

IEE Regulation C.10 offers the following alternatives (*a*) flameproof, pressurised or intrinsically safe apparatus, or apparatus giving an equivalent degree of protection, (*b*) normal apparatus at a height above the general floor level as determined by the authority responsible for safety requirements for the premises or (*c*) normal apparatus installed outside the area of risk.

In general areas, where a dangerous accumulation of petroleum vapour is unlikely to occur except at floor level, the appropriate authority will usually be satisfied with normal apparatus installed at least 1.3m above floor level.

Inspection pits, however, must be regarded as risk areas because the heavier-than air vapour spilled from petrol, or even vent pipes of vehicle fuel tanks, will tend to collect at the lowest level. Normal apparatus must never be installed; for example, pack-type fluorescent luminaires recessed into the pit wall and covered by a sheet of plain glass in a wooden frame are not satisfactory.

The Home Office Model Code for petrol filling stations recommends that pits smaller than 0.9m wide by 1.2m deep by 9m long should be treated as Zone 1 areas although larger pits in well-ventilated locations can be classed as Zone 2, except within 230mm of the pit floor, which should be Zone 1. Pits of any size which are located in Zone 2 areas must be classed as Zone 1.

Use of inspection lamps in pits is dangerous practice; the chances of explosion are equally great with extra-low voltage lamps as with mains voltage lamps. Provision of adequate fixed lighting of the appropriate type, and avoidance of installing socket-outlets in the vicinity, will promote pit safety.

Note: The terms Zone 1 and Zone 2 have replaced Division 1 and Division 2 but have the same meaning

6.10 Dusty environments

Far too often covers are left off electrical apparatus, or interiors are removed and left exposed, during installation and the equipment is not effectively cleaned before commissioning.

Utmost care should be taken to prevent the ingress of dust both at the construction stage — when building operations may result in clouds of dust or flying debris — and throughout the life of the installation.

Cable entry covers, trunking covers, unit doors and associated sealing strips should all be replaced with care; spare cable and conduit entries must be blanked off (Regulation B.95) and, where entries have been made too large, the excess space should be made good.

It is sometimes possible to avoid the risks associated with dust-laden atmospheres, and to avoid the cost of providing flameproof or dustproof apparatus, by segregating as much of the equipment as possible from the area in which the unfavourable conditions exist

When it is necessary to locate apparatus in permanently dusty environments, confirmation should be sought from the equipment manufacturer first that it satisfies the test requirements for exclusion of dust as set out in BS 3807 and, second, that surface temperatures are not likely to occur which could result in ignition of a dust deposit.

6.11 Static electricity

Electric shocks received from touching earthed metalwork are not usually caused by faulty electrical installations as the victims suspect.

Such shocks arise from the discharge of static electricity, generated by friction when people walk on non-conducting surfaces such as carpets containing nylon or other synthetic materials. The effect is particularly noticeable in centrally heated houses and carpeted offices where the humidity may be very low.

The effect can be minimised by ensuring that the humidity is maintained at a reasonable level. Where this is not possible, anti-static sprays are available for application to carpets and other surfaces which cause this effect. The spray is a surface treatment and must therefore be renewed at intervals dependent on the traffic across the treated surface.

6.12 Instantaneous water heaters

The use of instantaneous water heaters has become increasingly popular over recent years. In view of this, contractors ought to be aware of the need to notify the local Water Authority before commencing such an installation.

In many areas, local bye-laws require that at least seven days notice be given to the Water Authority of any alteration to an existing water supply.

This notice is necessary to enable the authority to offer advice and to assess whether the water supply is adequate. If found to insufficient, it many entail alterations to the water service before the installation could be approved.

7. INSPECTION and TESTING

7.1 Damage to solid-state controls

Dimmer switches, some modern cooker controls and other controls which incorporate solid-state devices, can be damaged during testing with a 500V insulation resistance tester.

There is increasing use of equipment with such sensitive electronic circuitry — dimmer switches, touch-operated switches, solid-state delay timers, photo-electric lighting switches, power controllers etc. — and care must be exercised when carrying out insulation resistance tests.

In commercial and industrial situations, where sensitive electronic equipment forms part of the fixed installation, a warning notice is frequently provided at the distribution board or supply intake position but this is unlikely to be the case where the above-mentioned equipment is installed.

To avoid damage to the components, it is recommended that a temporary link should be connected across the terminals of the device before making the tests of Regulation E.7 and E.8. Do not forget to remove all links at the conclusion of the test

7.2 Connection of socket-outlets

Incorrect connections at socket-outlets would be revealed if contractors included, in their final test and inspection procedure, an earth-loop impedance test at every socket-outlet.

Verification of polarity is essential for compliance with Regulation E.3 which also calls for a check that "wiring has been correctly connected to plugs and socket-outlets."

Incorrect polarity at socket-outlets is most likely to occur on installations wired in mineral insulated metal sheathed cable in which only black sleeving is used on the conductors.

When the supply has not been connected an earth-loop impedence test is not possible but, even then a continuity tester can be used to check polarity and to reveal other connection errors.

7.3 Open-circuit neutral connections

Insulation resistance tests should be conducted on all neutral conductors to locate any possible earth faults.

Such an earth fault may go unnoticed for a long period of time but if, as has happened on a number of occasions, the main neutral connection to the installation becomes open circuited, the whole neutral current for the installation will find its way to earth through the fault. This can result in severe overloading of the faulty conductor and subsequent fire risk.

7.4 Departures from the IEE Wiring Regulations

Customers sometimes have special installation requirements which involve departures from the IEE Wiring Regulations. Any such departures must be recorded on the completion certificate but, in addition, the contractor may have to decide whether a method has been competently specified and meets the spirit of the Regulations.

The introduction to the Regulations states that "any such departure or non-compliance should be the subject of the written specification of a competent body, or competent person or persons, and should result in a degree of safety from fire and shock not less than that assured by compliance with the Regulations."

An Assessment of New Techniques Scheme (ANTS) certificate is regarded as a written specification by a competent body. The use of a method or material covered by an ANTS certificate must be recorded on the completion certificate by entering the ANTS certificate number in the space provided.

Where the departure has been specified by another person, for example an architect, the contractor will have to decide whether it has been competently dealt with and produces the required degree of safety. Reference to Part 1 of the IEE Regulations will help show whether the proposed method complies with the spirit, if not the letter, of the Regulations.

If the contractor feels that the departure has not been competently specified, before undertaking the work he should advise the customer, in writing, of any possible danger. He should also state that a completion certificate will be issued on which all deviations will be recorded.

7.5 Inspection and consumers' appliances

Routine inspection of an installation should include a check that major appliances are safely installed, even though the inspection may not include portable appliances.

Cookers, washing machines, refrigerators and storage heaters are sometimes mounted on rollers which the user has bought separately from the appliance. This may make it possible for children to reach live connections from underneath, in which case it is in contravention of Regulation 1 and must be recorded on the inspection certificate. British Electrotechnical Approvals Board (BEAB) approval is no safeguard in this case because the rollers are not sold as part of the equipment.

Stability of the apparatus may also be affected by these rollers and, although not a direct concern of the IEE Wiring Regulations, is something to which the user's attention should be drawn.

7.6 Inspection of existing installations

It is frequently difficult to carry out a full inspection in occupied premises because, for example, the customer may not allow the removal of fitted carpets and lifting of floorboards for the examination of cable runs. It is important therefore, to record on the inspection certificate particulars of those parts of the installation that have been covered by, or excluded from, the inspection and test.

This information should be recorded in the blank space under "comments" in the inspection certificate prescribed in the IEE Regulations. Typical entries might read as follows: "This inspection did not include the wiring under floorboards"; "This certificate is based on an examination of the main switchgear, distribution board No. 7 and the final sub-circuits in Rooms 200-220 only"; or "It was not possible to remove lighting switches for examination because of damage to decorations."

In the absence of any statement to the contrary, it will be assumed that an inspection report relates to the entire installation and that every point has been examined for compliance with the Regulations.

In addition to the inspection certificate, it is often helpful if the customer can be given a schedule of circuits, and outlets connected to them; such a schedule will often help avoid misunderstanding if an order is subsequently placed for partial or full rewiring.

When making an inspection, care should also be taken to note any incorrect "do-it-yourself" extensions.

7.7 Inspection of "major alterations"

Regulation E.1 calls for inspection and testing upon completion of "every installation and every major alteration to an existing installation" but the precise definition of a "major alteration" is a matter for conjecture. A piece of work which would be regarded as a major alteration to an installation involving, say, six outlets would seem insignificant on an installation involving 6,000 outlets.

As a general guide, any addition or alteration to an existing installation which necessitates providing an additional final sub-circuit or an additional fuse-way, circuit-breaker way or other protective or control equipment, should be regarded as a major alteration and subjected to inspection and testing after completion.

However, it is in the contractor's own interest to inspect and test *any* installation work before handing it over to the consumer. A contractor may be held responsible for omitting to ensure that the equipment he has supplied and installed was safe for use. Fatalities have ocurred where this precaution was not taken

7.8 Existing installations affect alterations

It is essential, when inspecting and testing alterations to existing installations, to include those parts of the existing installation which affect the safety of new work, as required by Regulation E.1

This applies particularly to earthing systems: earthing leads may have become broken or disconnected, non-metallic inserts may have been installed in water pipes originally used for earthing, or earth-leakage circuit-breakers may have been wrongly installed or no longer operate satisfactorily

7.9 Annual inspection of filling stations

When conducting annual inspections of petrol filling station installations, as required by the Petroleum Regulation Acts, an electrical contractor is not responsible for determining the extent of alterations which are shown to be necessary.

The contractor is merely required to inspect, test and report upon the installation as it exists, drawing attention to deficiencies or departures from the requirements of the licensing authority. It is then the authority's responsibility to determine which matters, if any, require attention if continuance of the licence is to be granted.

In the case of obvious faults that can be rectified easily, such as a rusty fireman's switch, the contractor may use his discretion in taking the necessary remedial action.

7.10 Testing petrol pump circuits

Standard installation test equipment may be used with safety for testing petrol pump circuits when carrying out the annual inspection required by the Petroleum Regulation Acts, provided that the following test procedure is adopted. However, it is recommended that tests should not be made whilst petrol is being served to customers or delivered by tanker.

1. Before commencing the tests, remove the pump housing covers to allow any accumulated vapours to disperse, bearing in mind that petroleum vapour is heavier than air and may tend to persist at low level on calm days. As the interior of a pump housing is a Zone 1 area when fully assembled, the housing covers should not be replaced until conclusion of the following tests.

2. An earth-loop impedance test should be carried out, using a line-earth or neutral-earth tester as appropriate, as near as possible to the main switch controlling the installation. This will normally be located in a safe area. Additional earth-loop impedance tests should be made at socket-outlets in safe areas, but only the test figure at the main switch is required on the certificate.

3. Earth-continuity measurements are required between the consumer's earthing terminal and (i) each petrol pump and (ii) all other earthed metalwork. The measurements can be made with a standard battery-operated or hand-wound continuity tester.

4. Insulation resistance tests should be made in the normal way at the main switchgear and other points as may be necessary, after proving the effectiveness of earthing by the preceding tests.

7.11 Cross connection of ring circuits

Where ring final sub-circuits are installed, it is only too easy to make incorrect connections especially when a number of 2.5 sq mm cables are awaiting connection. The fact that such errors occur underlines the importance of the test for ring circuit continuity to Regulation E.10.

To make this test effectively, it is essential that *all* pairs of ring circuit conductors are disconnected and separated before applying a continuity tester; if only one pairs of cables is disconnected at a time, cross-connection of ring circuits will not be revealed.

Examples of the type of error which can occur include the outgoing leg of one ring connected to the return leg of another ring or a water heater circuit cable connected with one leg of a ring to a 30A way and the other leg taken to a 15A way.

In each case, two circuits are interlinked and fed from two sources with the result that the removal of a fuse or opening of a miniature circuit-breaker does not isolate either circuit because both circuits are still fed from the other protective device.

7.12 Installation testing

On completion of an installation the following tests, listed in the note at the beginning of Section E of the IEE Wiring Regulations, must be undertaken:

1. Verification of polarity. To check that the polarity of the supply to the installation is correct and that all single-pole fuses, circuit-breakers, switches and thermostats are in the live conductor, that plugs and socket-outlets are correctly connected and that the outer contacts of Edison-type screw lampholders are connected to the neutral conductor (Regulation E.2).

2. Tests of effectiveness of earthing. To check (i) that the impedance of earth-continuity conductors and bonding leads does not exceed the permitted value prescribed in Appendix 6 (Regulation E.3) and (ii) that the earth-loop impedance is satisfactory (Regulation E.4) and (iii) that earth-leakage circuit-breakers operate correctly (Regulation E.5).
3. Insulation resistance tests. To measure (i) the insulation resistance of all conductors to earth (Regulation E.7) and (ii) the insulation resistance between conductors (Regulation E.8).
4. Test of ring circuit continuity. To ensure that the live, neutral and earth-continuity conductors of ring circuits are continuous and connected to the correct terminals at the origin of the ring (Regulation E.10).

If any of these tests reveals a fault, the preceding tests should be repeated once the fault has been cleared.

The importance of correct inspection, testing and certification cannot be overstressed; in the case of a periodic inspection, it is also essential to indicate on an inspection certificate whether any parts of an installaton were not accessible for visual inspection and therefore could only be tested by the use of instruments.

7.13 Completion and inspection certificates

From the wording of Regulations E.11 and E.13 and also note 2 following Regulation E.12, it is clear that the IEE expects completion and inspection certificates to be signed by, or on behalf of, the contractor responsible for constructing an installation or by making an inspection.

The contractor may, of course, authorise another competent person to sign certificates on his behalf, in which case the name and address of the contractor must appear on the certificate to make it clear that the person signing is doing so on behalf of the contractor.

7.14 Period between inspections

While all installations should be regularly inspected and tested, in some cases, there is a statutory requirement for inspections at specified intervals but in others, the period is left to the discretion of the designer or installer.

In the following list of typical installations, the period between inspections is coded "M" where mandatory and "R" where recommended.

It must be remembered, however, that recommended intervals may become mandatory if incorporated in Local Authority conditions of licence for various types of premises.

Type of installation	*Reference*	*Period between inspections*	
General installations	IEE Regs pp 85, 87	5 years or less	R
Farms	,, K.39	3 years	R
Caravans	,, J.23	1 year	R
Construction sites — temporary installations	,, H.3	3 months	R
Fire alarms	BS CP1019	1 year*	R
Emergency lighting	BS5266: Part 1	not specified*	R
Cinemas, etc.	SI 1959 No. 1530 clause 27	1 year	M
Petrol filling stations	Local Authority Conditions of Licence	1 year	M
Launderettes	GLC General Powers Act and similar legislation in other areas	1 year	M
Churches	"Lighting & Wiring of Churches 1972"	2 years (new installations)	R
		1 year (installations over 5 years old)	R
Factories	Electricity (Factories Act) Special Regulations	regular intervals	M

* Other intervals are recommended for tests of operation and of batteries and generators.

8. INSTALLATION CHECK LISTS

When an NICEIC inspecting engineer examines an electrical installation, he hopes to find full compliance with the IEE Wiring Regulations which consist of 405 individual regulations, many of them containing multiple requirements.

The following lists show, with regulation numbers, some of the principal items for which the inspecting engineers have learned to look. They provide a useful guide to the factors requiring particular attention when installing and inspecting equipment.

8.1 Wiring accessories

General (applicable to all types of accessory)

Complies with BS 846 or other appropriate Standard (F.1)
Box or other enclosure securely fixed (1(b))
Edge of flush box level with wall surface to ensure non-combustible enclosure (B.69)
No sharp edges on cable entries, screw heads etc. which could cause mechanical damage to cables (B.25, B.27)
Non-sheathed cables, and cores of cables from which sheath has been removed, not exposed outside the non-combustible enclosure (B.69)
Conductors correctly identified (B.54)
Bare earth-continuity conductors sleeved green/yellow (or green alone up to December 1977) (D.28, D.29, B.54)
Terminals tight and containing all strands of the conductors (B.61)
Cable clamp correctly used or clips fitted to cables to prevent mechanical strain on the terminals (B.61)
No excessive removal of insulation (B.64)
Adequate current rating (C.3)
Suitable for the conditions likely to be encountered (C.4)
Protected against damp (C.6)
Not exposed to water, oil, steam, vapour or mechanical damage (C.7)
Metal box or other enclosure earthed (D.3)

Lighting switches

Complies with BS 3676 (F.1)
Readily accessible (A.59)
Single-pole switches connected only in live conductors (A.8)
Correct colour coding of switch wires as live conductors (B.54)
Earth-continuity conductor terminated in an earthing terminal (sheathed wiring) (D.7)
Earthing of exposed metalwork, e.g. metal switch plate (D.3)
Earthing connection to metal switch grid if left loose for depth adjustment when flush mounted (D.3)
Out of reach of a person using a bath or shower (D.17) or immediately outside the door (A.61)
Adequate current rating (C.3)
Suitable for inductive circuits or de-rated where necessary (C.2, G.2)
Out of reach of wash troughs etc. on agricultural and horticultural installations (K.32)

Labelled to indicate purpose, where this is not obvious (A.14)

Ceiling roses

Complies with BS 67 (F.1)
Live terminal shrouded to prevent accidental contact (C.19)
Earth-continuity conductor connected to earthing terminal (C.21)
Not more than one flex unless designed for multiple pendants (C.20)
Flex support devices used (B.33(xi), B.61)
Switch wires identified as live conductors (B.54)
Holes in ceiling above rose made good to prevent spread of fire (B.40)
Not connected to a supply exceeding 250V (C.18)
Suitable for the mass suspended (C.27, B.132)

Socket-outlets

Complies with BS 196, BS 546, BS 1363 or BS 4343 (C.29, F.7)
Mounting height not less than 150mm above the floor or working surface. (C.30 Note 2)
Correct polarity (C.30)
Not installed in a bathroom or shower room (D.18)
Not within 2.5m of a shower cubicle in a room other than a bathroom (D.18)
Controlled by a switch, on agricultural and horticultural installations (K.7) and where the supply is direct current (A.57)
Connected to the same phase in any one room except as permitted by the exemption to Regulation A.20 (A.20)
Protected from water where mounted on a floor (C.30 Note 1)
Not used to supply a water heater having uninsulated elements (C.50)
Not supplied by an auto-transformer (C.32)
Earth-continuity conductor connected directly to the earthing terminal of the socket-outlet, on a sheathed wiring installation (D.5)
Earthing tail from the earthing terminal of the socket-outlet to a metal box having two sliding lugs on a sheathed installation (D.3)
Earthing tail from the earthed metal box, on a conduit installation, to the earthing terminal of the socket-outlet (D.5)

Joint box

Accessible for inspection (B.73)
Protected against mechanical damage (B.74)
Earth-continuity conductors correctly connected (B.61)
Earth-continuity conductors not less than 1.5 sq mm if connected outside the box (D.28(v))

Fused connection box

Out of reach of a person using a bath or shower (D.17)

Cooker control unit

Within 2m of cooking appliance(s) (A.29)
Readily accessible (A.59)
Cable sizes adequate for the current demand of the cooker (A.27)

8.2 Conduits

General

Completely erected before drawing-in of cables (B.88)
Securely fixed and adequately protected from mechanical damage (B.89)
Inspection fittings accessible (B.90)
Permitted number of cables not exceeded (B.91)
Solid elbows and tees used only as permitted (B.93)
Ends of conduit reamed and bushed (B.94)
Adequate boxes (B.95)
Unused entries blanked off (B.95)
Not containing non-electrical pipes or tubes (B.53)
Self-ventilating and provided with drainage holes (B.96)

Rigid metal conduit

Complies with BS 31, BS 2706 or BS 4568 Parts 1 and 2 (B.97)
Earthed (B.98, D.3)
Segregated from, or bonded to, extraneous fixed metalwork (B.53, D.14)
Mechanically and electrically continuous (B.98)
Satisfactory earth continuity (D.28(iv), D.29(v))
Cables correctly rated, allowing for grouping, ambient temperature, class of excess-current protection and length of run—see item **1.4** (B.20)
Live and neutral cables bunched in the same conduit (B.100)
Heavy gauge and corrosion-resistant in damp situations (B.42, B.125, B.126, B.127, K.16)
Conduit not used as an earth-continuity conductor underground (B.125)
3m maximum span between buildings (B.126)
Not used for spanning gaps between buildings on agricultural and horticultural installations (K.22)
Not used as sole earth-continuity conductor where accessible to livestock (K.29)

Rigid non-metallic conduit

Complies with BS 4607 Parts 1 or 2 (B.101)
Ambient and working temperatures within permitted limits (B.102)
Provision for expansion and contraction (B.104)
Mass of luminaire suspended from box does not exceed 3kg (B.70, B.103)
Temperature of box used for suspension of luminaire does not exceed 60degC (B.70, B.103)
Cables rated at currents appropriate to p.v.c-insulated cables even where higher grades of insulation are used (B.102 Notes).

Flexible metal conduit

Complies with BS 731 Part 1 (B.97, B.106)
Not used as sole earth-continuity conductor (B.107)
Adequately supported (B.108)

8.3 Trunking

General

Securely fixed and adequately protected against mechanical damage (B.110)
Constructed of metal or non-combustible insulating material (B.111)
Covers securely fixed (B.113)
Entries protected against ingress of water (B.113)
Permitted number of cables (space factor 45 per cent) not exceeded (B.118)
Not containing non-electrical services (B.53)
Barriers to prevent excessive temperature rise in vertical trunking (B.35)
Barriers to prevent spread of fire (B.40)
Holes surrounding trunking made good (B.40)
Category 2 circuits partitioned from Category 1 circuits or insulated for the highest voltage present (B.49)
Category 3 circuits partitioned from Category 1 circuits or wired in mineral-insulated metal-sheated cables (B.49)
Common outlets for category 1, 2 and 3 circuits provided with screens, barriers or partitions (B.50)
Cables correctly rated, allowing for grouping, ambient temperature, class of excess current protection and length of run (B.20)

Metal trunking

Live and neutral cables bunched in the same metal trunking (B.112)
Protected against damp or corrosion (B.42)
Earthed (D.3)
Joints mechanically sound and electrically continuous where used as an earth-continuity conductor (D.28(iv), D.29(v))

8.4 Insulated cables

Non-flexible cables

Correct type (B.1)
Correct voltage rating (B.19)
Correct current rating (B.20)
Permitted volt-drop not exceeded (B.23)
Diversity correctly applied (B.20, B.24, A.21, A.22)
Protected against mechanical damage and abrasion (B.25, B.26, B.27)
PVC insulated and/or sheathed cables not installed in refrigerated spaces (B.7)
Non-sheather cables protected by enclosure in conduit, duct or trunking (B.28)
Not exposed to direct sunlight or, if so exposed, of a suitable type (B.29)
Not run in lift shaft unless part of the lift installation and of the permitted type (B.30)

Correctly selected and installed for use underground (B.125)
Correctly selected and installed for use on exterior walls etc. (B.126)
Correctly selected and installed for use overhead (B.127)
Internal radii of bends in accordance with Table B.1M (B.32)
Correctly supported (B.33)
Selected for ambient temperature (B.34)
Selected for maximum temperature of insulation (B.36)
Selected for explosion protection (B.39)
Not exposed to water etc. unless suitable for such exposure (B.41, B.42, B.43, B.44)
Metal sheaths and armour earthed (B.85, D.3)
Metal sheaths and armour segregated from, or bonded to, extraneous fixed metalwork (B.53, D.14)
Terminations identified (B.54)
Terminations mechanically and electrically sound (B.60)
All wires securely contained in terminals etc., without mechanical strain (B.61)
Termination of aluminium cables correctly made (B.62, B.63, B.77)
No excessive removal of insulation (B.64)
Braid, sheath etc. cut back to prevent leakage (B.65)
Enclosure of terminations (B.69)
Use of flameproof enclosures where necessary (B.71)
Glands correctly selected and fitted (B.72)
Joints mechanically and electrically sound and accessible for inspection, except as permitted otherwise (B.73)
Joints adequately insulated (B.74)
Earthed concentric wiring (including c.n.e. cables) used only as permitted (B.119–124)
Not used for connection of movable apparatus (B.128)

Flexible cables and cords

Correct type (B.4, B.12)
Correct voltage rating (B.19)
Correct current rating (B.20)
Sheathed (and armoured if necessary) where exposed to mechanical damage (B.13)
Suitably sheathed where exposed to contact with water (B.15) and oil or petrol (B.16)
Sheathed where used for final connections to fixed apparatus etc. (B.130)
Selected for resistance to damage by heat (B.17, B.34, B.36)
Segregation of Category 1 and 2 circuits (B.51)
Category 3 circuits not contained in common cable with Category 1 circuits (B.52)
Cores correctly coloured throughout (B.57)
Prohibited core colours not used (B.58)
Joints effected by a cable coupler (B.78, F.10)
Armouring not used as sole earth-continuity conductor (B.129)
Not used as fixed wiring except as permitted (B.130)
Final connections to fixed apparatus as short as possible and connected as stated (B.131)
Mass supported by pendants not exceeding values stated (B.132)
Cross-sectional area of core used as an earth-continuity conductor not less than

that of the associated current-carrying conductor (D.28(ii))

8.5 Earth-continuity conductors

(Items marked by an asterisk also apply to earthing leads and bonding leads)

Cables incorporating earth-continuity conductors comply with the relevant B.S. (D.28(i))
Cross-sectional area of e.c.c. in a flexible cable or cord not less than that of the largest associated current-carrying conductor (D.28(ii))
Cable sheaths and/or armouring used as an e.c.c. comply with the relevant B.S. (D.28(iii))
Joints in metal conduit, duct or trunking comply with Regulation D.29(v) (D.28(iv))
Flexible conduit not used as a sole e.c.c.. (D.28(iv))
*Cross-sectional area of other copper conductors in accordance with Table D.2M (D.28(v))
*Resistance of e.c.c.s other than copper not more than the prescribed copper conductor (D.28(v))
*Copper conductors, other than strip, of 6 sq mm or less protected by insulation D.29(i))
*E.C.C. at termination of sheathed cables insulated with sleeving D.29(ii))
*Bare e.c.c. protected against mechanical damage and corrosion (D.29(iii))
*Insulation, sleeving and terminations identified by colour green-and-yellow (or green only up to December 1977) (B.54, B.55, B.57, D.29(iv))
*Joints mechanically and electrically sound, electrically continuous and protected against corrosion (D.29(v))
*Connection to cable sheaths and/or armouring and pipes effected by soldered joints or clamps (D.29(vi))
Copperclad aluminium conductors not used as bonding leads (D.29(vii))
Pipes and structural metalwork not used as e.c.c. (D.30)

9. USEFUL PUBLICATIONS

Designers and installers of electrical equipment need to refer to numerous publications which range from those giving guidance on good installation practice to those laying down statutory conditions which must be observed. The following list is intended to assist readers in determining the source of supply of the documents referred to elsewhere in this book, together with other relevant publications.

British Standards and British Standard Codes of Practice

Details of all publications are contained in the British Standards Institution Year Book, published annually. Electrical standards and codes of practice are listed in Sectional List SL26, available free of charge.

Orders by post to British Standards Institution,
Sales Branch,
101 Pentonville Road,
London N1 9ND.

Counter service for personal callers at British Standards Institution, 2 Park Street, London W1A 2BS, and at regional offices in Birmingham, Bristol, Dundee, Glasgow, Leeds, Liverpool and Norwich.

Statutory Instruments and other Government Publications

Orders by post to Her Majesty's Stationery Office,
P.O. Box 569,
London SE1 9NH.

Bookshop service at 49 High Holborn, London WC1V 6HB, and Government bookshops at Belfast, Birmingham, Bristol, Cardiff, Edinburgh and Manchester. Also obtainable through booksellers.

Electrical Research Association (ERA) Reports

Obtainable from Electrical Research Association
Cleeve Road,
Leatherhead,
Surrey KT22 7SA.

IEE Wiring Regulations and ANTS Certificates

Orders by post to Instution of Electrical Engineers,
P.O. Box 26,
Hitchin,
Herts SG5 1RS.

Counter service for personal callers at 2 Savoy Place, London WC2R 0BL.

Miscellaneous Publications

Obtainable by post from the addresses given:

"Code of Practice for Electrical Safety in Beer Dispense in Licensed Premises." Brewing Publications Ltd, 42 Portman Square, London W1H 0BB.

"Lighting and Wiring of Churches." Church Information Office, Dean's Yard, London SW1 3NZ.

"Construction Safety." Electricity, Section 10. National Federation of Building Trades Employers, 82 New Cavendish Street, London W1M 8AD.

"The Installation and Operational Aspects of Private Generating Plant. G26." Distribution Engineering Branch, The Electricity Council, 30 Millbank, London SW1P 4RD.

"Codes of Practice re. Principle of Construction, Electrical Requirements, Model Licence Conditions etc." Association for Petroleum and Explosives Admistration, c/o J. S. Crewe Esq., Consumer Protection Department, Bedfordshire County Council, 23 Gadsby Street, Bedford MK40 3HP.

"Registered House-Builders Handbook." National House-Building Council, 58 Portland Place, London W1N 4BU.

"The IES Code — Interior Lighting." Lighting Division, The Chartered Institution of Building Services, 49 Cadogan Square, London SW1X 0JB.

INDEX BY SUBJECT

Adaptors, conduit **2.19**
Adjustable lugs **3.3**
Air ducts **3.8**
Alarm systems, fire **1.16**
Alterations **2.22, 7.7, 7.8**
Anti-static spray **6.11**
Appliances, fixed **1.6, 1.7**
Armouring, cable **3.13**
Assessment of New Techniques Scheme (ANTS) **5.8, 7.4**

Bathrooms **1.11, 6.1, 6.3, 6.5, 6.6**
Baths **3.11, 5.1, 6.1, 6.3**
Batteries, alarm system **1.16**
Bell transformers **1.9**
Boilers, gas **1.10**
Bonding **3.8, 3.9, 3.10, 3.11, 3.12, 3.13, 3.16, 3.17**
Borrowed neutrals **2.11**
Bushes **2.19, 3.5**

Cable,
- armouring **3.13, 3.18**
- combined neutral and earth (c.n.e.) **2.21**
- damage to **2.1, 2.2, 2.3, 2.4, 2.25, 2.26**
- mechanical protection **2.1, 2.2, 2.3 2.26, 2.27**
- mineral insulated **2.4, 2.5, 8.4**
- multi-core **2.6, 8.4,**
- split concentric **2.21**

Cable sizes **1.1, 1.2, 1.4, 1.5, 1.7, 1.20, 8.4**
Cable tray **3.2**
Ceiling rose **8.1**
Ceilings, cable in **2.1, 2.25**
Cellars, cable in **2.2**
Central heating boiler **1.10**
Circuit-breaker,
- miniature and moulded-case **4.6**
- earth-leakage **3.14, 4.7, 4.8, 4.9, 4.10, 4.11, 4.12, 4.13, 4.14, 6.2**

Clamps, bonding and earthing **3.10**
Class of protection **4.5**
Completion certificate **7.4, 7.13**
Conduit,
- general, **1.23, 2.17, 2.18, 2.20, 3.5, 3.19, 8.2**
- fittings **2.10, 2.19, 3.5, 5.8**
- flexible **8.2**
- metallic **1.23, 2.17, 8.2**
- non-metallic **1.12, 3.6, 5.8, 8.2**

Confederation for the Registration of Gas Installers (CORGI) **1.10**
Connection units **1.7, 3.4, 8.1**
Connector blocks, moulded **2.10**
Connectors, crimp or compression **2.9**
Consumer's control unit **1.1, 4.3**
Control gear, lighting **2.5, 5.9**
Cooker control unit **3.4, 6.3, 6.7, 8.1**
Cookers, split level **6.7**
Cord grips **2.8**
Corrosion **2.17**
Current rating,
- apparatus **1.7, 1.8, 4.3, 4.4, 5.11**
- cables **1.1, 1.2, 1.4, 1.5, 1.7, 1.17, 2.4**

Damp environments **5.1, 6.1, 6.2, 6.3, 6.4, 6.5, 6.6, 6.8**
Damp-proofing **4.11**
Departures from Regulations **7.4**
Dimmer switches **7.1**
Direct contact **4.7**
Discharge lighting **1.3, 2.5, 5.10, 5.11**

Discrimination, e.l.c.b.s **4.9**
Distribution boards **1.1, 1.2, 2.12, 4.3**
Ducting **2.13**
Dust **6.10**
Earth continuity **2.15, 2.16, 2.17, 2.18, 2.19, 3.5, 3.8, 3.20, 7.12, 8.5**
Earth continuity conductor **1.23, 3.2, 3.6, 3.7, 3.13, 8.5**
Earth electrode **3.14, 4.12, 4.13, 4.14**
Earth-leakage circuit-breaker:
  all types **3.14, 4.8, 4.12, 4.15**
  current-operated **4.7, 4.12, 4.14, 6.2**
  voltage-operated **4.9, 4.10, 4.11, 4.12, 4.13**
Earth-loop impedance test **2.7, 7.2 7.10, 7.12**
Earthing,
  cable **3.18**
  equipment **3.1, 3.2, 3.3, 3.4, 3.19**
  leads **3.13, 3.15, 4.12**
  responsibility **3.21**
  tails **3.3, 3.4**
  terminals **3.1, 3.3, 3.4, 3.21**
Electricaire **1.8, 3.8**
Electrical Research Association **1.5**
Electronic devices **7.1**
Emergency lighting **1.15**
Excess-current protection **1.1, 1.2, 4.4, 4.5**
Extension leads **2.23**
Extension work **3.14, 3.17, 4.12, 7.6, 7.7, 7.8**
Extract fans **1.11**
Extra-low voltage **1.17, 1.18**
Explosive atmospheres **1.3, 6.9**

Factories **2.18, 7.14**
Fans, extract **1.11**
Farms **1.23, 2.2, 6.8, 7.14, 8.2**
Fire alarms **1.16, 3.20**
Fire prevention **1.22**
Fireman's switch **1.3**
Flameproof equipment **6.9, 6.10**
Floating cores **2.6**
Floorboards **2.1**
Fuses **4.4, 4.5**

Garages **6.9, 7.9**
Gas boilers **1.10**
Gas pipes **3.9, 3.10**
Glands, **3.5, 3.18**

Harmonics **5.10**
Health and Safety at Work etc. Act **2.23**
Heaters,
  radiant **1.3, 1.6**
  thermal storage **1.6, 1.7, 4.3, 4.10**
  water **1.3, 1.6, 6.12**

Indirect contact **4.7**
Inspection **7.5, 7.6, 7.7, 7.8, 7.9, 7.12, 7.13, 7.14**
Inspection pits **6.9**
Insulation resistance test **7.1, 7.3, 7.12**
Interconnections, switchgear **1.2**
Intrinsic safety **6.9**
Isolation **1.3, 1.8, 1.10, 1.11, 1.13, 1.15, 4.1, 4.2, 4.16, 5.12**

Joint boxes **5.6, 8.1**
Joints,
  cable **3.7**
  conduit or trunking **2.15, 2.16**
Joists **2.1**

Kitchens **6.3**

Labels **1.10, 1.16, 1.19, 3.10, 4.12, 6.3**
Lampholders **5.1, 5.2, 6.4**
Lighting,
  discharge **1.3, 2.5, 5.10, 5.11, 5.12**
  emergency **1.15**
  fluorescent **4.16, 5.7, 5.8, 5.9, 5.10, 5.11**
  three-phase **5.10**
  tungsten **1.22, 4.16, 5.6**
Lighting track **5.4, 5.5**
Lighting trunking **5.3**
Luminaires – see under "Lighting"
Locknuts **3.5**
Loft insulation **2.3**

Main switch **1.1, 4.1**
Making good **2.24**
Meter tails **1.1, 1.2**
Mineral insulated cable **2.4, 2.5, 7.2**
Miniature circuit-breaker **4.6**
Motors **1.13, 1.14, 2.18**
Multi-gang switches **5.10**

Neutral conductors **2.11, 2.12, 5.10, 7.3**

Paint 3.5
Parallel earth paths **4.9**
Petrol filling stations **6.9, 7.9, 7.14**
Petrol pumps **1.3, 7.10**
Petroleum Regulation Acts **7.9, 7.10**
Pits, inspection **6.9**
Polarity **2.7, 7.2, 7.12**
Polystyrene **2.3**
Power factor **5.11**
Pressurised equipment **6.9**
Prospective fault current **4.6, 4.7**
Protection,
  direct and indirect contact **4.7**
  earth leakage **3.21, 4.7, 4.8, 4.9, 4.10, 4.11, 4.12, 4.13, 4.14, 6.2**
  excess current **4.5, 4.6**
Protective multiple earthing **3.13, 3.16, 4.15**

Rating factors **1.4, 5.11**
Rating tables **1.4, 1.5**
Rats **2.2**
Ring final sub-circuits **1.6, 3.6, 7.11, 7.12**
Rising mains **4.2**
Rivets **2.16**
Rollers **7.5**
Roof void **2.2, 2.3, 2.26**

Sauna baths **6.1**
Segregation **1.15**
Service fuse **1.1**
Service stations **6.9, 7.9**
Shaver points **6.4, 6.5, 6.6**
Sheathed wiring **3.3, 3.6**
Shop fitting **5.9**
Showers **5.1, 6.4, 6.12**
Sinks **3.11, 3.12**
Skirting trunking **2.14**
Socket-outlets **2.7, 3.3, 3.4, 6.2, 6.3, 6.4, 7.2, 8.1**
Socket-outlets, e.l.v. **1.18**
Solid-state controls **7.1**
Split-level cookers **6.7**
Static electricity **6.11**
Stop buttons **1.13, 1.14**
Storage heaters **1.6, 1.7, 4.3, 4.10**
Supply authority's fuse **1.1, 4.1**
Surge voltage **2.5**
Swimming pools **6.2**
Switch,
  double-pole **1.3**
  fireman's **1.3**
  isolating **1.3, 1.8, 1.10, 1.11, 1.15, 4.1, 4.16, 5.12**
  lighting **4.16, 5.10, 8.1**
  main **1.16, 4.1, 4.2**
Switchboards **1.2, 1.21**

Temperature **1.12, 2.4, 2.10, 5.6, 5.7, 5.8**
Temporary wiring **2.22, 2.23, 2.28, 7.14**
Terminations **2.6, 2.7, 2.8, 2.9, 3.18, 5.6**
Testing **2.7, 4.8, 7.1, 7.2, 7.3, 7.10, 7.11, 7.12, 7.14**
Three-phase supplies **5.10**
Time switch **3.1**
Toilets **1.11**
Transformers **1.3, 1.9**
Track lighting **5.4, 5.5**
Trunking **1.4, 2.13, 2.14, 2.15, 2.16, 5.3, 8.3**

Underground wiring **1.23, 2.20, 2.21**

Ventilation fans **1.11**
Vermin **2.2**
Volt drop **1.4**
Voltage,
  exceeding 250V **1.19**
  extra-low **1.17, 1.18, 1.19**
Voltage surges **2.5**

Water heater **1.3, 1.6, 6.12**
Water pipes **3.9, 3.10, 3.14, 4.14**

INDEX BY REGULATION NUMBER

This index covers Regulations referred to in the text, but not in the installation check list of Chapter 8.

A.1 **1.1, 1.16**
A.3 **1.1, 1.2**
A.4 **4.1**
A.6 **1.16**
A.9 **1.3, 1.17**
A.10 **1.2, 1.20**
A.11 **1.2, 4.2**
A.12 **1.2, 4.2**
A.15 **2.12**
A.17 **1.19**
A.18 **5.10**
A.19 **1.19**
A.20 **5.4**
A.24 **2.11, 2.12, 5.10**
A.25 **1.17**
A.26 **1.17, 5.4**
A.27 **6.7**
A.29 **6.7**
A.32 **1.6**
A.36 **3.6**
A.43–55 **1.17**
A.52 **3.6**
A.58 **1.3, 1.9, 1.11, 1.15 4.16**
A.59 **4.16**
A.60 **4.16**
A.61 **6.1**
A.62 **1.7, 1.15**
A.63 **1.13, 1.14**
A.64 **1.13, 1.14**

B.25 **2.25, 2.26, 2.27, 4.12**
B.26 **2.1, 2.26**
B.28 **2.13, 2.14, 5.9**
B.33 **2.25, 2.26, 2.28, 3.12**
B.36 **5.7**
B.42 **2.17**
B.45–52 **2.14**
B.53 **2.25**
B.54 **4.12**
B.60 **1.22**
B.61 **2.8, 2.9, 3.18**
B.69 **1.9, 1.22, 2.13, 2.14, 5.5, 5.9 5.10**
B.70 **5.8**
B.73 **2.9**
B.74 **2.10**
B.78 **1.22**
B.89 **2.18**
B.95 **6.10**
B.96 **2.17**
B.97-100 **1.23**
B.98 **3.19**
B.100 **5.10**
B.102 **1.12, 2.20**
B.103 **1.12**
B.104 **1.12, 2.20**
B.111 **2.14**
B.119–124 **2.21**
B.125 **2.20, 2.21**
B.130 **5.9**

C.3 **4.3**
C.6 **6.8**
C.10 **6.9**
C.11-16 **1.22**
C.23 **5.1**
C.24 **5.2**
C.31 **1.3**
C.50 **1.3**

D.2 **3.6**
D.3 **1.23, 2.20, 3.19, 3.2, 3.3**
D.4 **3.1, 3.8**
D.5 **3.4**
D.8 **1.9, 1.17**
D.10-14 **3.16, 3.17**
D.12 **3.9**
D.13 **3.10,**
D.14 **3.8, 3.10, 3.11 3.17**
D.15–19 **6.1**
D.16 **5.1, 6.2, 6.4**
D.17 **6.2, 6.3, 6.4**
D.18 **6.2, 6.3, 6.4**
D.19 **6.6**
D.20 **4.7**
D.22 **3.21, 4.7, 4.15**
D.23 **3.21, 4.7**
D.24 **4.12, 4.15**
D.25 **4.12**
D.28 **3.7, 3.13, 3.18**
D.29 **2.15, 2.18, 2.19, 3.5, 3.7, 4.12, 4.13**
D.31 **4.12, 4.13**
D.32 **4.12, 4.13**
D.33 **3.10, 4.12, 4.13**
D.35 **3.14, 4.14**

E.1 **4.12, 7.7, 7.8**
E.2 **7.2, 7.12**
E.3 **3.20, 7.2, 7.12**
E.4 **7.12**
E.5 **4.8, 7.12**
E.7 **7.1, 7.12**
E.8 **7.1, 7.12**
E.10 **7.11, 7.12**
E.11 **6.3, 7.13**
E.12 **7.13**
E.13 **7.13**

F.1 **1.8**
F.7 **6.5**

G.3 **5.11**
G.5 **5.9**
G.11 **1.3, 5:12**
G.12 **5.12**
G.13 **1.3**

H.1-3 **2.23, 2.28**
H.3 **7.14**

J.17 **1.22**
J.18 **1.22**
J.23 **7.14**

K.20 **1.23**
K.39 **7.14**

NOTES

NOTES

NOTES

NOTES

NOTES

NOTES

NOTES